PHYSICS RESEARCH AND TECHNOLOGY

NEW DEVELOPMENTS IN QUANTUM OPTICS RESEARCH

PHYSICS RESEARCH AND TECHNOLOGY

Additional books in this series can be found on Nova's website under the Series tab.

Additional e-books in this series can be found on Nova's website under the e-book tab.

PHYSICS RESEARCH AND TECHNOLOGY

NEW DEVELOPMENTS IN QUANTUM OPTICS RESEARCH

NICOLE STEWART
EDITOR

nova publishers
New York

Library of Congress Cataloging-in-Publication Data

New developments in quantum optics research / editor, Nicole Stewart.
pages cm -- (Physics research and technology)
Includes bibliographical references and index.
ISBN 978-1-63483-394-3 (hardcover) -- ISBN 1-63483-394-5 (hardcover) 1. Quantum optics--Research. I. Stewart, Nicole, 1971- editor. II. Series: Physics research and technology.
QC446.2.N476 2015
535'.15--dc23

2015025942

Published by Nova Science Publishers, Inc. †New York

CONTENTS

PREFACE

This book focuses on new developments in quantum optics research. Chapters include computer-aided design (CAD) flow for large-scale quantum circuits; the effects of cooperation between the Stokes and anti-Stokes modes in Raman scattering processes and an analysis of the different properties, such as entanglement and quantum degree of polarization for two-mode states of the radiation field which is of great importance for a deeper understanding of the correlations that occur in the system.

A computer-aided design (CAD) flow for large-scale quantum circuits is presented in Chapter 1. The proposed flow along with other stand-alone tools are implemented in a tool suite called Next-generation Quantum Computing Systems (NQCS) toolbox. This toolbox receives as input a physical machine description of a quantum computing fabric, a quantum error correction (QEC) code, a quantum control (QC) protocol, as well as a quantum algorithm, and produces as output physical resource requirements and a machine control language specification of the quantum algorithm, protected by the given QEC code and the QC protocol. The NQCS toolbox comprises of several stand-alone tools to perform compilation, logic synthesis, and physical optimization steps. The automated CAD flow is made possible by creating a set of well-defined interfaces between different levels of design abstraction and developing tools that operate efficiently and reliably within any abstraction levels. This modular design approach simplifies the task of developing and ensuring the correctness of the overall design flow. By taking advantage of hierarchical nature of the quantum circuit specifications, NQCS is able to handle large-scale quantum benchmarks while producing high-quality results.

Chapter 2 proposes the cooperative excitation and absorption of three cavity modes which corresponds to pump, Stokes and anti-Stokes photons stimulated by the atomic inversion. The three modes collective Roman emission is defined introducing the cooperative description between photon of cavity photons. In the case, when the scattering rates in the Stokes and anti-Stokes modes coincide, the SU(2) and SU(1,1) symmetries are applied for a simple description of these cooperative processes. The possibilities to realize this effect in the free space is proposed, replacing the cavity modes with dipole active excited atoms in tow-quantum interaction with dipole-forbidden transitions of D atom. The statistical properties and detection method are proposed using the information entropy and atomic correlation functions.

In Chapter 3, the authors investigate two quantities that characterize the two-mode states of the radiation field: quantum degree of polarization and entanglement. The authors evaluate analytically the quantum degree of polarization for the three-photon Bell-diagonal states by exploiting two distance-based measures: Hilbert-Schmidt and Bures. Two examples are considered further in our analysis: the Werner state and a family of Bell-diagonal states defined by a parameter. For each example they compare the concurrence with the two types of quantum degrees of polarization mentioned above.

In: New Developments in Quantum Optics ... ISBN: 978-1-63483-394-3
Editor: Nicole Stewart, pp. 1-50

Chapter 1

COMPUTER-AIDED DESIGN FOR NEXT-GENERATION QUANTUM COMPUTING SYSTEMS

Alireza Shafaei, Mohammad Javad Dousti* and Massoud Pedram
Department of Electrical Engineering,
University of Southern California, Los Angeles, CA, US

Abstract

A computer-aided design (CAD) flow for large-scale quantum circuits is presented in this chapter. The proposed flow along with other stand-alone tools are implemented in a tool suite called Next-generation Quantum Computing Systems (NQCS) toolbox. This toolbox receives as input a physical machine description of a quantum computing fabric, a quantum error correction (QEC) code, a quantum control (QC) protocol, as well as a quantum algorithm, and produces as output physical resource requirements and a machine control language specification of the quantum algorithm, protected by the given QEC code and the QC protocol. The NQCS toolbox comprises of several stand-alone tools to perform compilation, logic synthesis, and physical optimization steps. The automated CAD flow is made possible by creating a set of well-defined interfaces between different levels of design abstraction and developing tools that operate efficiently and reliably within any abstraction levels. This modular design approach simplifies the task of developing and ensuring the

*E-mail address: dousti@usc.edu

correctness of the overall design flow. By taking advantage of hierarchical nature of the quantum circuit specifications, NQCS is able to handle large-scale quantum benchmarks while producing high-quality results.

PACS: 05.45-a, 52.35.Mw, 96.50.Fm

Keywords: Physical design, quantum computing, compiler, fault-tolerant quantum computation, quantum mapper

1. Introduction

Quantum computing offers the potential of dramatically higher performance on certain problems compared to classical computing. As a result, substantial research efforts have focused on discovering high-level algorithms that could operate on a quantum computer [1, 2, 3, 4], and developing a wide variety of physical technologies that might make the construction of a quantum computer possible (see [5] for an overview of various quantum technologies). However, less attention has been paid to the *computer-aided design* (CAD) aspects of quantum computing which include compilation and mapping of a quantum algorithm onto a physical machine. This is the focus and contribution of this chapter.

One way to view the mapping of a quantum algorithm on a quantum computer is as a quantum physics experiment that is under the control of a classical computer [5]. The experimental apparatus consists of several parts: (i) A *quantum core* contains the physical qubits whose quantum-mechanical properties are used to accomplish the computation; (ii) Classical qubit-level control and detection circuitries serve to initialize, manipulate, and read out the state of the quantum core; (iii) A *real-time classical computer* directs the quantum experiment by issuing a sequence of instructions to the control and readout electronics, to be performed on the quantum core. The instruction sequence is generated from a *high-level quantum program* by an off-line compilation process. Similarly, result verification and output post-processing are also done off-line.

A realistic quantum computer is subject to *noise* and faces numerous limitations and constraints. For example, components of a quantum core are sensitive to environmental disturbances and to errors in the control and detection systems. If ignored, either of these effects can rapidly lead to computational error (see below). In addition, the technology used to implement the quantum computer will be subject to design constraints on parallelism, geometry, quantum channel width, etc., that further narrow the options for implementing quantum algorithms.

There is strong theoretical evidence that arbitrarily long quantum computations could be carried out on realistic quantum computers if the error rates per gate and per time step could be reduced below a level known as the "fault-tolerant accuracy threshold" [6]. This process, known as *fault-tolerant quantum computation*, relies in part on the ability of *quantum error-correcting codes* to protect the quantum information used in the computation from random disturbances and the quantum decoherence effect.

Using currently known quantum technologies, implementation of fault-tolerant quantum computation requires massive computational resources. More precisely, realizing a quantum algorithm with relatively small input size tend to result in a very large number of qubits and quantum operations. Therefore, it is essential to develop an automated design flow and associated toolbox to deal with the complexity of implementing a quantum algorithm on a quantum computing fabric. Motivated by this consideration, this work proposes a CAD flow for quantum information processing that receives as input a *physical machine description* (PMD), a quantum error correction (QEC) code, a *quantum control* (QC) protocol, as well as a quantum algorithm, and produces different artifacts including but not limited to a *machine control language* (MCL) code, and physical resource requirements. The proposed CAD flow is called - (NQCS). This flow along with other stand-alone tools are implemented in a tool suite which will be referred to as the NQCS toolbox in the rest of the chapter.

The NQCS toolbox comprises of several stand-alone tools to perform required tasks. These tools interact with each other through well-defined interfaces. This feature allows the tool developers to work on individual tools independently, and any new tool can be simply added to the whole flow in the future. Moreover, NQCS users have the flexibility of augmenting the interface files when and if needed. To finish, this modular design approach simplifies the task of ensuring the correctness of the overall design flow.

As a proof of concept, the NQCS toolbox uses *Scaffold Compiler* [7] as the quantum compilation tool to generate HF-QASM of the given quantum algorithm, and QUFD [8] to provide the latency and MCL of fault-tolerant gates for the given PMD, QEC, and QC. Furthermore, Squash 2 [9] is adopted as the quantum physical design tool which translates the HF-QASM to the final circuit MCL, and reports the final physical resources.

The remainder of this chapter is organized as follows. A list of key definitions required for the rest of this chapter is provided in Section 2. NQCS requirements are explicitly described in Section 3. Section 4 analyzes functional requirements and provides detailed discussions on the key components in the design flow. Interface specifications are presented in Section 5. Verification methods and performance measures are explained in Section 6 and Section 7, respectively. Finally, recommendations for further reading are presented in Section 8.

2. Key Definitions

The following terms have been used throughout this chapter.

- (PMD). A description for the underlying quantum physical machine technology, which include the qubit de-coherence super-operator, control response functions, inter-qubit connectivity, geometric constraints (e.g., nearest neighbor interactions), qubit movement constraints, crosstalk, and detector parameters.

- (QC). A protocol to control the quantum dynamics of the quantum hardware.

Native Instruction. A quantum gate that belongs to the instruction set of a PMD. (e.g., R_x, R_y, and R_z, which are equal to the rotation of the quantum bit around the X-, Y-, and Z-axes by an arbitrary angle).

Basic Operations. Quantum gates that are used in fault-tolerant quantum computation. The standard universal set of the basic operations includes H, T, CNOT, S, X, Y, and Z gates. Quantum circuits are built out of basic operations. However, to realize a given quantum circuit in the given PMD, each basic operation should be implemented by using native instructions (e.g., $T = R_z(\pi/4)$ and $H = R_y(\pi/2)R_z(\pi)$, up to a global phase).

- (QGS). A universal set of quantum gates to be used in a fault tolerant quantum computation. The standard universal set includes T, S, H, X, Y, Z, and CNOT. Basic single-qubit measurements of X and Z are also included in the QGS.

Primitive Cells. PMD of a quantum circuit fabric consists of a 2D array of identical *primitive cells* (or cell for short). Each cell contains some sites for generating/initializing qubits, reading them out, performing operations on one or two physical qubits, and resources for moving information about qubits.

Tile. Dealing with the 2D array of primitive cells is very cumbersome and unwieldy. So in practice we build another 2D array of super-cells (which we call *tiles*). Each tile comprises of an $n \times n$ array of primitive cells. We can thus deal with mapping a quantum algorithm to this tiled architecture. An example of a 2×2 tiled architecture where each tile consists of a 4×4 cell is shown in Figure 7. The layout consists of tiles where each tile stores a single logical qubit and ancillae to enable all single-qubit logical gates. To execute a CNOT gate (as a two-qubit operation), target and control logical qubits should initially be collocated.

- **(ULB).** ULB is analogous to a Configurable Logic Block in an FPGA device, which means it can implement any of a set of target functions.

- **(CULB).** A ULB that is capable of performing one- and two-input fault-tolerant gate operations on one or two logical qubits. It can also perform quantum error correction on a single logical qubit.

- **(MULB).** A ULB that can store several idle logical qubits. It periodically performs QEC on the logical qubits (one after another) in order to maintain their fidelity.

Scaffold. A high-level programing language that is used to describe a quantum algorithm. More information about the Scaffold and its associated compiler can be found in [10] ans [7], respectively.

- **(QASM).** A gate-level programming language for describing a quantum circuit [11].

- **(HF-QASM).** An intermediate representation between the front-end and back-end parts of the NQCS toolbox. This representation is a variant of QASM with hierarchy added and gate types restricted to fault-tolerant gates.

- **(XML).** XML is a simple and powerful language, which has been used extensively to design textual databases. It is a markup language that defines a set of rules for encoding documents in a format that is both human-readable and machine-readable. The design rules are specified with XML schema files. There are many XML development resources and solutions, XML parsers, etc., which are quite helpful for us.

Mini Languages. A set of XML-based specifications that describe the PMD, the QC protocol, and the - (QEC) code. To describe the encoding/decoding circuity of the QEC code, a quantum circuit description language (e.g. QASM) is also required.

Machine Description and Protocol Libraries. After processing Mini languages, the machine description and protocol libraries are created, which include: (i) the realization of each gate in QGS using native instructions of the given PMD by considering the QC protocol, and (ii) encoding/decoding circuits for the given QEC code.

- (QIDG). Graph representation of a quantum circuit, in which nodes represent (assembly) instructions and edges capture data dependency between operands of the instructions. This graph is used in scheduling, placement and routing steps of the quantum physical designer.

- (QUFD). A tool designed by NQCS team that generates a library of MCL codes for fault-tolerant quantum gates.

- (QPD). Main tool of the back-end which generates MCL of the quantum circuit through scheduling, placement, and routing.

- (MCL). A low-level textual language which encodes simple control instructions that would be delivered to the classical real-time computer. The set of instructions is therefore necessarily tightly coupled to the physical machine description. The classical computer would then translate the machine control language instructions into technology-dependent commands (e.g., lasers, detectors, magnets, RF-signal generators, etc) for execution by the specific hardware components that control the qubits and quantum gates. The machine control language may include items such as qubit movement primitives, readout primitives, control primitives, qubit addressing, and pulse shapes. The MCL is analogous to the classical - (ISA).

- **(RC).** A major task in the toolbox that calculates the physical resources (i.e., the number of physical qubits, the number of physical gate operations, and circuit delay) needed to implement a quantum algorithm.

- **(MCT).** A Toffoli gate with multiple controls.

- **(EFFBD).** An EFFBD represents the behavior for a system or components of a system. The purpose of the EFFBD is to indicate the sequential relationship of all functions that must be accomplished by a system. EFFBDs show the same tasks identified through functional decomposition and display them in their logical, sequential relationship at different levels of a hierarchy.

- **(SAT).** In computer science, SAT is the problem of determining if the variables of a given Boolean formula can be assigned in such a way as to make the formula evaluate to TRUE. Equally important is to determine whether no such assignments exist, which would imply that the function expressed by the formula is identically FALSE for all possible variable assignments. In this latter case, we would say that the function is unsatisfiable; otherwise it is satisfiable. SAT was the first known example of an NP-complete problem.

3. NQCS Toolbox Requirements

The NQCS toolbox requirements can be divided into functional and interface requirements as discussed below.

3.1. Functional Requirements

The NQCS toolbox provides the following functional requirements:

1. The NQCS toolbox allows the quantum programmer to write the quantum algorithm in Scaffold language providing quantum data types, quantum operations, classical operations, and appropriate control flow structures. Scaffold is discussed in [10].

2. The NQCS toolbox enables

   (a) PMD expert to describe a new PMD architecture, along with error rates and latencies for various quantum instructions,

(b) QEC expert to describe a new quantum error correction code, and
(c) QC expert to describe a new quantum control protocol.

The above items are provided by the use of Mini languages.

3. The NQCS toolbox provides an efficient quantum compiler for the quantum programmer. Quantum compiler is discussed in [7].

4. The NQCS toolbox generates the machine description language (i.e., MCL code) for the quantum programmer given the Scaffold program, QC, QEC code, and PMD specification. MCL generation is discussed in Section 4.5.

5. The NQCS toolbox includes standalone tools for calculating physical resources for the given Scaffold program, QC protocol, QEC code, and PMD specification. Resource calculation is discussed in Section 4.6.

6. The NQCS toolbox provides a set of software utilities such as a timing simulator and the NQCS quantum programming environment. Timing simulator is discussed in Section 4.5.1.

7. The NQCS toolbox supports a mechanism to analyze the correctness of algorithm implementations in Scaffold. Verification is discussed in Section 6.

3.2. Interface Requirements

The NQCS toolbox provides the following interface requirements:

1. The input to the whole design flow is a quantum algorithm written in Scaffold [10].

2. The input to the front-end tool suite is a program in Scaffold, and the output is HF-QASM.

3. The input to the quantum physical design (back-end) is HF-QASM, and the output is MCL code.

4. Inputs to the resource calculation tool are HF-QASM as well as error rate and delay information extracted from protocol libraries, and the output is physical resources (# of physical qubits, # of physical gates and their types, # of clock cycles).

5. The input to the Mini language processor is a set of XML and QASM files for describing (PMD, QC, QEC), and the output is protocol libraries.
6. The input to the library designer is a set of protocol libraries, and outputs are MCLs for gates, tile design and the 2D title architecture.
7. The NQCS toolbox provides an integrated development environment, in the sense that a quantum algorithm written by a quantum programmer in Scaffold will be transformed into the final MCL after applying several automated tools including a quantum compiler, a synthesizer, etc.

Interface requirements are verified by monitoring and validating (as much as possible) inputs and outputs of the front-end and back-end tools, resource calculator, Mini language processor, and library designer.

Based on the above interface requirements, the following stand-alone tools are available in the NQCS toolbox:

- Quantum compiler
- Mini language processor
- Reversible logic synthesizer
- Physical designer
- Verifier(s)
- Integrated Development Environmnet (IDE) and visualizers

4. Functional Requirements Analysis and Specification

Functional analysis is the process of identifying, describing, and relating the functions a system must perform in order to fulfill its goals and objectives. Functional analysis is logically structured as a top-down hierarchical decomposition of those functions. Several techniques are available to do functional analysis. The primary functional analysis technique is the *functional flow block diagram* technique. These diagrams show the network of actions that lead to the fulfillment of a function. This section uses - (EFFBD) representation, which is based on [12, Appendix F], to depict functional decomposition of the NQCS toolbox.

4.1. - (EFFBD)

The purpose of the EFFBD is to indicate the sequential relationship of all functions that must be accomplished by a system. EFFBDs show the same tasks identified through functional decomposition and display them in their logical, sequential relationship at different levels of a hierarchy. For example, the entire tool chain flow of a quantum circuit designer can be defined in a top level EFFBD, as shown in Figure 1. Each block in the first level diagram can then be expanded to a series of functions, as shown in the second level diagrams in Figure 3 to Figure 13. EFFBDs use the following symbols and notation:

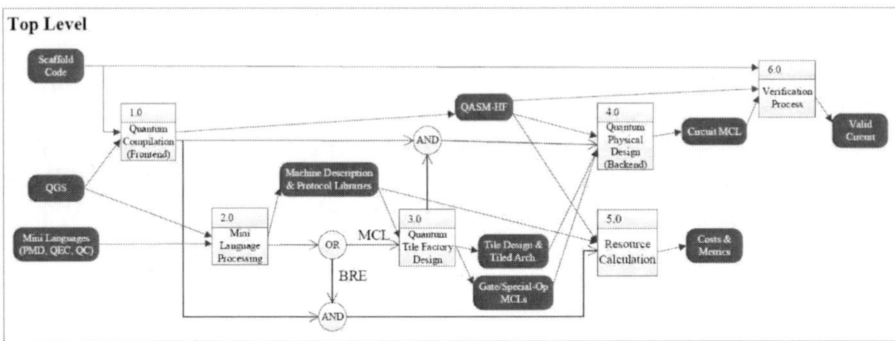

Figure 1. EFFBD of NQCS top level. For EFFBD notation please refer to Section 4.1.

- Functions are represented as rectangular boxes (blocks) with an associated label number. Functional (control) flow is shown by head-flat arrows, whereas data flow overlay to capture data dependencies is represented by head-filled arrows and elongated ovals.
- Each function receives a unique label number that can be used as identification, and is shown in the top-left corner of the function. The numbering scheme establishes identification and relationships that carry through all the diagrams and facilitate traceability from the lower levels to the top level.
- Lines connecting functions indicate functional flow and not lapsed time or intermediate activity.
- Diagrams are laid out so that the flow direction is generally from left to right. The diagrams show both input (e.g., “Mini language Processing”) and output (e.g., “Quantum Physical Design”), thus facilitating the definition of interfaces and control processes.

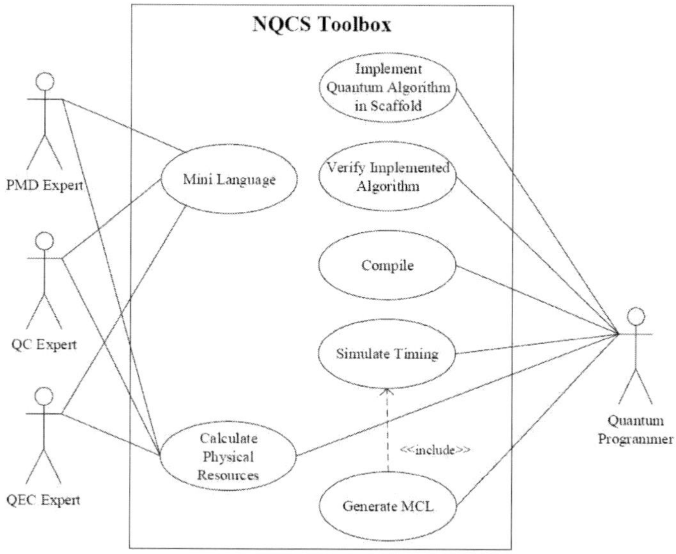

Figure 2. Use-case diagram of the NQCS toolbox.

- Each diagram contains a reference to other functional diagrams to facilitate movement between pages of the diagrams.
- Gates are also used, including “AND”, “OR”, “Go or No-Go”, sometimes with enhanced functionality, including exclusive OR gate (XOR), iteration (IT), repetition (RP), or loop (LP).
- A circle is used to denote a summing gate when AND/OR is present. As expected, AND indicates parallel functions and all conditions must be satisfied to proceed (i.e., concurrency) whereas OR indicates that alternative paths can be satisfied to proceed (i.e., selection).
- “Go or No-Go” decision blocks are shown by a diamond. The “Go” path from the decision block is executed if the condition is met; otherwise, the “No-Go” path is executed. These symbols are placed adjacent to lines leaving a particular function to indicate alternative paths.

An EFFBD specification of a system is complete enough that it is executable as a discrete event model, capable of dynamic as well as static validation. EFFBDs provide freedom to use either control constructs or data triggers or both to specify execution conditions for the system functions.

4.2. The NQCS Toolbox

The top- (first-) level diagram of the NQCS toolbox is depicted in Figure 1, which is then expanded to second-level diagrams in Figure 3 through Figure 13. NQCS, at the top-level view, performs the following two major tasks:

- **Design automation** for a quantum compiler, which maps a quantum algorithm into an MCL code, and
- **Resource calculation**, which reports the physical resources that are sufficient to implement a quantum algorithm on a target PMD.

These tasks consider QEC to protect qubits against noise, and QC to increase the fidelity. As a result, a combination of quantum algorithm, PMD, QEC code, and QC protocol acts as the input to each task. More accurately, the primary inputs to NQCS are classified as follows:

- A quantum algorithm written in Scaffold [10].
- Mini languages, which are a set of XML-based specifications that describe the PMD, the QC protocol, and the QEC code.
- A universal set of quantum gates to be used in a fault tolerant quantum computation (previously referred to as basic operations.) This universal set is referred to as QGS and includes T, S, H, X, Y, Z, and CNOT [6]. For the given QEC code, the exact implementation of each of the gates in the QGS is given as a QASM file. We also include basic single-qubit measurements of X and Z in the QGS, since they are a necessary part of a fault-tolerant quantum computing scheme. In addition to the standards gates, new gates can also be added to this set as long as an efficient, fault-tolerant implementation for the new gates exists in the given QEC code, and implementation costs are properly considered in the synthesis step.

To perform quantum computation fault-tolerantly, quantum gates should be chosen from the QGS to build quantum circuits. To this end, NQCS adopts the following approach (cf. Figure 1):

1. "1.0 Quantum Compilation (Front-end)" in Figure 1, receives Scaffold code and the QGS, compiles the code, and synthesizes it to a circuit consisting of gates from QGS. This intermediate representation is called HF-QASM.

2. The HF-QASM representation is then translated by "4.0 Quantum Physical Design (Back-end)" to a physical quantum circuit language, i.e., the MCL description. However, prior to the physical design, an MCL code for each gate in the QGS is generated as follows:

   (a) After processing Mini languages (cf. function 2.0 in Figure 1), the *machine description and protocol libraries* are created, which include: (i) the realization of each gate in QGS using native instructions of the given PMD by considering the QC protocol, and (ii) encoding/decoding circuits for the given QEC code.

   (b) "3.0 Quantum Library Design" uses the data set generated in (a) in order to produce the internal design of a tile, the 2D tiled-architecture of the PMD, and MCL codes for QGS members and also the *special operations* (or, special-ops for short). Special-ops are defined in this chapter as the *hop operation* (i.e., one step move operation in up, down, left or right directions), and *syndrome extraction* for the QEC code.

3. In addition to the back-end, which generates the MCL code, "5.0 Resource Calculation" computes the *costs and metrics* for the quantum algorithm that include the number of physical qubits, number of physical gate operations, and the circuit delay (i.e., number of physical time steps to execute the quantum algorithm).

4. Verification shows proof of compliance with toolbox requirements where validation shows that the toolbox accomplishes its major tasks. "6.0 Verification Process" handles these cases and will be discussed in Section 6.

Based on the above discussion, it can be seen that the NQCS uses a *meet-in-the-middle approach* to generate the MCL code. On one hand, the top-down development in the front-end tool suite breaks the Scaffold code into gates that are chosen from the QGS. On the other hand, the bottom-up development in the library designer tool generates optimized MCL codes (in terms of gate or qubit counts) for each gate of the QGS. These two paths meet each other in the back-end, where high-level gates are scheduled, placed and routed on the 2D tiled architecture, and then each high-level gate as well as move operation are replaced by the appropriate MCL code in order to build the circuit MCL. Note that the path that goes through Mini language processing and quantum library design tools is an offline process that executes independently of the quantum algorithm.

A key benefit of the meet-in-the-middle approach is that the efficiency of the top-down (forward) synthesis approach and the accuracy of the bottom-up physical design approach are simultaneously realized. Additionally, the design tool scalability is ensured by hiding from the algorithm and code developers the physical design details of implementing logic operations (and moves) in a given PMD for given QC and QEC protocols. Ease of verification of the design by separating front-end transformations from gate library verification is another benefit.

The NQCS toolbox relies on a set of well-defined *use-cases*, whose purpose is to capture system functions that affect the interface between the system and the outside world. Well-written use-cases offer the analysis, development, testing, and documentation teams an invaluable guidebook. Precisely, a use-case is a formalized story that describes how someone procedurally interacts with a proposed software system. In the rest of this section, we develop use-case diagrams for various components in the NQCS toolbox. These components are those that an expert quantum programmer/researcher can utilize in order to write a given quantum algorithm and subsequently transform the written program into a machine-level code for a given combination of PMD, QEC, and QC. The rest of this section provides detailed descriptions for functions of the top-level diagram along with the appropriate use-cases.

4.3. Front-end

A given quantum algorithm is written in Scaffold [10]. The specification of Scaffold allows two kinds of modules for describing a quantum algorithm: (1) *normal modules* which are written with quantum variables and gates, along with classical or quantum control structures, and (2) *Classical-to-Quantum-Gate* (-(CTQG)) modules. CTQG modules are a key feature of Scaffold that allows a programmer to express a desired quantum functionality using classical programming. Rather than viewing everything in terms of quantum gates, CTQG

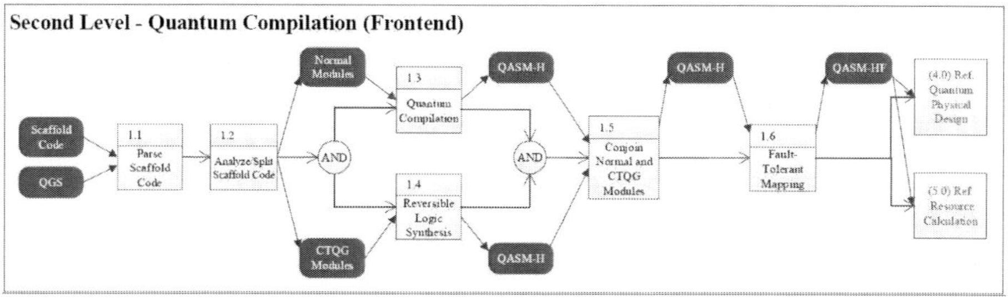

Figure 3. EFFBD of NQCS front-end.

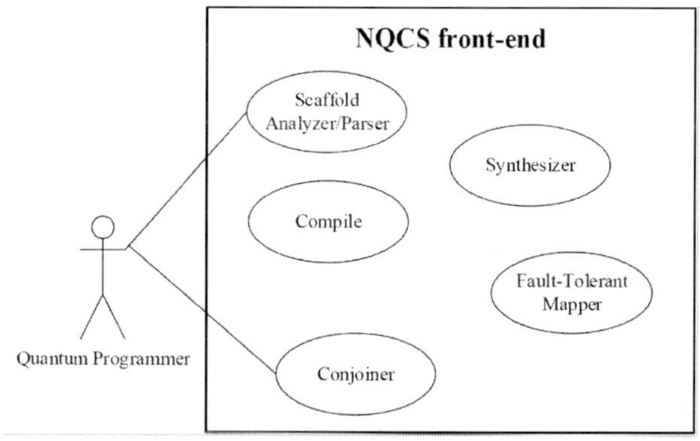

Figure 4. NQCS front-end use-case diagram.

modules allow programmers to describe the functionality of some parts of the algorithm from a higher perspective. They are written as classical C-like code, with the intent of being synthesized to reversible logical circuits that are compatible with a quantum circuit. Normal and CTQG modules are suitable for specifying the quantum and oracle parts of the algorithm, respectively. These two parts are distinguishable by the Scaffold grammar. In addition to the quantum and classical modules, there are various control structures that allow the programmer to identify the control flow of Scaffold code. More details are available in Table 1.

After parsing the Scaffold code (cf. function 1.1 of Figure 3), function 1.2 of Figure 3 first analyzes the Scaffold code, identifies normal and CTQG modules, and forwards them along with the required control structures through the appropriate path in the design flow to the corresponding compilation/synthesis tools. CTQG modules are synthesized by a *reversible logic synthesis* tool (cf.

Table 1. The detailed description for the "Implement Quantum Algorithm in Scaffold" use-case

Use-case name	**Implement Quantum Algorithm in Scaffold**
Related requirements	Requirement 1 in Section 3
Goal in context	The NQCS toolbox provides an integrated development environment to specify a quantum algorithm and subsequently transform and optimize the quantum program into the machine-level code. Scaffold is the NQCS quantum programming language.
Preconditions	Must have a detailed specification of the algorithm, specifying the different parts in quantum circuit or high-level (e.g., oracle, etc.) flow.
Successful end conditions	A new Scaffold program is developed.
Fail end conditions	-
Primary actors	Quantum Programmer
Secondary actors	-
Trigger	The quantum programmer starts writing the Scaffold program.
Main flow	Action
	(Step #1) Quantum Programmer studies the algorithm in detail. (Step #2) Quantum Programmer decides on which parts to specify using classical code and which parts to write at quantum gate level. (Step #3) Quantum Programmer uses the available language features and syntax to implement the specifications. (Step #4) Quantum Programmer must be aware of the decision's implications and possible implementation costs (e.g., large non-reversible logics that use a lot of ancilla to be made reversible, arbitrary rotations, etc.)

function 1.4 of Figure 3), e.g., - (RMDDS) [13]. The normal modules are fed into the *quantum compilation* tool (cf. function 1.3 of Figure 3), which executes in parallel to the reversible synthesis path. The tool is responsible for synthesizing quantum logic gates into QGS. In the first step (not shown in the figure), the quantum compiler unrolls the loops, evaluates conditions, and interprets the gate sequence and control path described by the programmer. In the second step (again not shown), the quantum compiler implements arbitrary rotations in terms of QGS gates by using the Solovay-Kitaev algorithm [14]. In this case, a sequence of machine instructions must be generated to approximate the arbitrary rotation to the required precision. Analysis in [15] shows that static code generation can, in some cases, lead to a terabyte of code to realize needed rotations. Further, some rotation angles are unknown until the runtime, requiring dynamic code generation. Dynamic code generation, however, can result in significant tradeoffs in terms of the execution time overhead versus code quality (code size and rotation precision). Moreover, dynamic code generation will be performed on classical (non-quantum) computing resources, which may or may not have a clock speed advantage over the target quantum technology.[1] In [15], the design space formed by these trade-offs of dynamic versus static code generation is explored. They also introduced several techniques to provide smoother trade-offs for dynamic code generation and evaluate the viability of options in the design space. Note that this step produces complex quantum gates such as multiple-control Toffoli gate. The output of the second step is - (H-QASM) (see Section 5).

Compiled normal and CTQG modules are merged together by a conjoiner tool (cf. function 1.5 in Figure 3). The *conjoiner* is responsible for putting the codes together and resolving hybrid, nested instantiations, viz., instantiation of a classical module inside a quantum. To manage the size of output, the conjoiner avoids flattening the output and maintains the hierarchical structure (instantiation of modules inside the other modules). Circuit optimizations, similar to [16] and [17], may also be performed by the conjoiner tool to further optimize the quantum circuit. Finally, a *fault-tolerant mapping* tool (cf. function 1.6 in Figure 3) takes the output of the conjoiner, and maps it to gates in the QGS. The output of the fault-tolerant mapper is the HF-QASM representation. Based on the above discussion, the frond-end use-case diagram is shown in Figure 4, and the front-end actions are summarized in Table 2.

[1]For example, although modern classical processors run at gigahertz clock speeds, operations on trapped ions run at kilohertz speeds, whereas superconducting qubits run at gigahertz speeds.

Table 2. The detailed description for the "Compile" use-case

Use-case name	**Compile**
Related requirements	Requirement 3 in Section 3
Goal in context	The Scaffold program is compiled into HF-QASM representation.
Preconditions	The Scaffold program should pass the verification step.
Successful end conditions	A new HF-QASM is generated.
Fail end conditions	Compiler generates a list of errors including the location and type of the error.
Primary actors	Quantum Programmer
Secondary actors	-
Trigger	The quantum programmer asks the NQCS to compile the Scaffold program.
Main flow	Action
	(Step #1) The Quantum Programmer asks the NQCS toolbox to compile the Scaffold program. (Step #2) The Scaffold program is parsed, and split into normal and CTQG modules (cf. function 1.2 in Figure 3). (Step #3) CTQG modules are synthesized to H-QASM using a reversible logic synthesis tool. (Step #4) Normal modules are compiled into H-QASM using the Quantum Compiler. (Step #5) CTQG and normal H-QASM files are conjoined to produce the final H-QASM representation. (Step #6) The H-QASM representation is transformed into the HF-QASM representation. (Step #7) HF-QASM serves as input to other tools including the quantum physical design tool or the resource calculation tool.

4.4. Library Designer

The purpose of the library designer is to design the internal structure of a tile and the 2D tiled architecture, and to generate a library of gate models.[2] The library designer comprises of two stand-alone tools:

1. The *Mini language processor*, shown in Figure 5: The input to the Mini language processor is a set of XML-based "Mini languages", which describe the PMD, the QC protocol, and the QEC code. The Mini Language processor produces a QASM description file for each gate of the QGS using the native instructions of the PMD.

2. The *quantum tile factory designer*, shown in Figure 6: This tool implements each gate of the QGS in a fault-tolerant manner, attaches the syndrome extraction circuit to the output(s) of the gate, and places the resultant quantum circuit into the architectural tile. The MCL code for the QGS gate being fully realized in the tile is then generated.

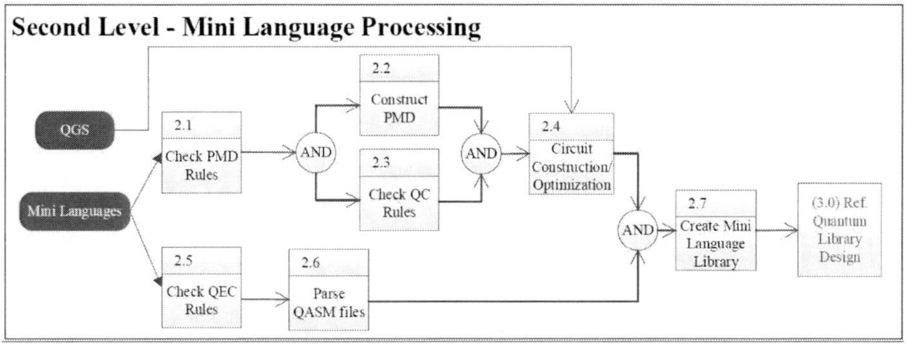

Figure 5. EFFBD of NQCS Mini language processor.

After appropriate high-level transformations, the quantum circuit is represented as a QIDG (function 3.1 in Figure 6), in which nodes represent (assembly) instructions and edges capture data dependency between operands of the instructions. To capture the quantum no-cloning theorem, which forbids fan-out in quantum circuits, a pre-processing step is implemented on the QIDG to

[2]This model includes MCL code, delay, and error rate for each QGS, move operation, and MCL code for applying syndrome extraction.

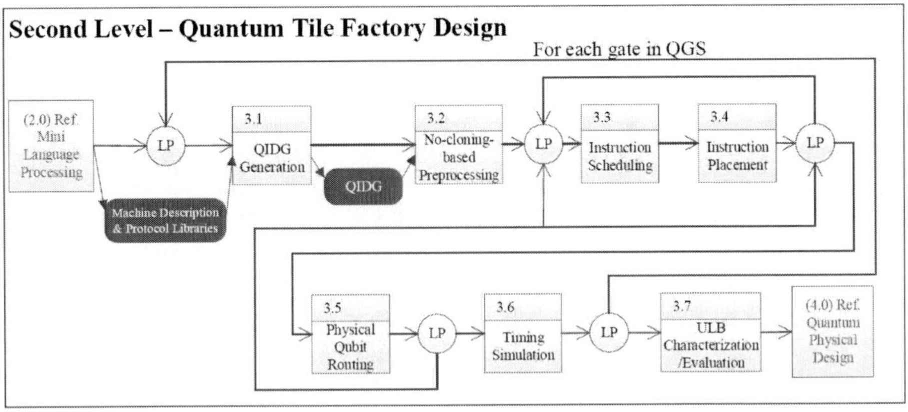

Figure 6. EFFBD of NQCS quantum tile factory designer. LP denotes a loop in the EFFBD.

resolve the multiple read dependencies between instructions (function 3.2 in Figure 6). In the next step, the instructions are scheduled (function 3.3 in Figure 6) and placed (function 3.4 in Figure 6) on the quantum fabric, and the qubits are routed (function 3.5 in Figure 6). Scheduling, placement, and routing problems in the context of quantum computing are defined as:

- **Scheduling**: Given a QIDG, the scheduling problem is to reduce the resource pressure (e.g., the number of concurrent instructions) while minimizing the total latency of the circuit.
- **Placement**: Given a total ordering of the instructions on the circuit (as the output of the instruction scheduling step), the placement problem is to assign physical (logical) qubits and instructions (operations) to the cells (tiles) such that the communication time (or total latency of the computation) is minimized.
- **Routing**: Given the position of each qubit and the instruction schedule, the qubit routing problem is to minimize the total routing time.

In the last step, the optimal size of the tile is determined such that various fault-tolerant realizations of required quantum instructions can be implemented in any tile in the fabric with acceptable latency and without any resource waste. We refer to this tile as - (ULB). When this optimal tile size is known, the tool generates the MCL code for each gate of the QGS. For circuits coded with block QEC, the proposed algorithms have been implemented as a quantum - (CAD) tool called - (QUFD) [8]. Timing simulation (cf. function 3.6 of Figure 6) is used in QUFD tool to verify the generated MCL code.

The quantum tile factory designer is responsible for designing two types of high-level (architectural) ULBs.

- **- (CULB)**: This tile is capable of performing fault-tolerant gate operations on one or two logical qubits. It can also perform quantum error correction on a single logical qubit. The type of error correction is specified as an input by the QEC code.
- **- (MULB)**: This tile can store several idle logical qubits. It periodically performs QEC on the logical qubits (one after another) in order to maintain their fidelity. This is based on the fact that idle qubits need to be refreshed approximately once every 100 operation cycles in concatenated codes [18]. Therefore, applying QEC on idle qubits in a round-robin manner is adequate.

To achieve a regular tiled architecture for the front-end tool suite, width and height of the MULB should be the same as those of the CULB. Hence, the number of idle logical qubits that the MULB can store is a function of CULB size, idle qubit de-coherence rate, and the delay of applying a QEC on a logical qubit. An example of the tiled architecture for the superconducting PMD is shown in Figure 7.

4.4.1. Mini Language Processor

Specifying PMD, QC and QEC in a simple language format provides the ability to modify/extend the existing PMD, QC and QEC protocols. It also provides the flexibility to capture new PMDs, QC, or QEC protocols even after the design of the quantum toolbox. Considering these objectives, we need to devise a descriptive language for PMD, QC and QEC to be used in this process. The designed language should be easy to use by the PMD, QC and QEC developers (Figure 8 and Table 3). Yet, it should be expressive, concise, and robust. For these reasons, we have opted to use the XML language to specify the PMD, QC and QEC protocols. In addition to XML, some type of quantum circuit description language is needed to specify the required QEC protocols viz. to provide a

Figure 7. A 2 × 2 tile-based architecture in superconducting PMD, where each tile is a 4 × 4 cell.

description of the encoding and decoding circuitry. QASM or Scaffold may be used for this purpose.

The definition of a PMD is based on primitive cells that are specific to that PMD. The quantum fabric can be generated by replicating these primitive cells and considering links between the neighboring cells. In addition to these layout definitions, the set of native instructions that can be implemented in each of the PMDs is specified in an XML file. For the QC definition language, native instructions of each PMD are replaced with a series of native and control instructions to increase their fidelity. These replacements are defined using an XML file. To capture different QEC schemes (block vs. topological), we suggest different definition languages for different QEC schemes. These definition languages include the basic parameters specific to each coding scheme. In addition to these parameters, circuit implementation of basic operations, ancilla generation and parity check for different coding schemes are also specified using augmented QASM files. Moreover, the NQCS toolbox considers two different paths for applying concatenated and topological QEC to a quantum circuit.[3]

[3]The proposed mini-language processor can easily handle any modifications to the encoding/decoding circuitry for a supported QEC - regardless of whether it is concatenated or topological code. On the other hand, handling a new (and previously unseen) is a very challenging task - something that requires further research in order to assess its feasibility and if so, ways of

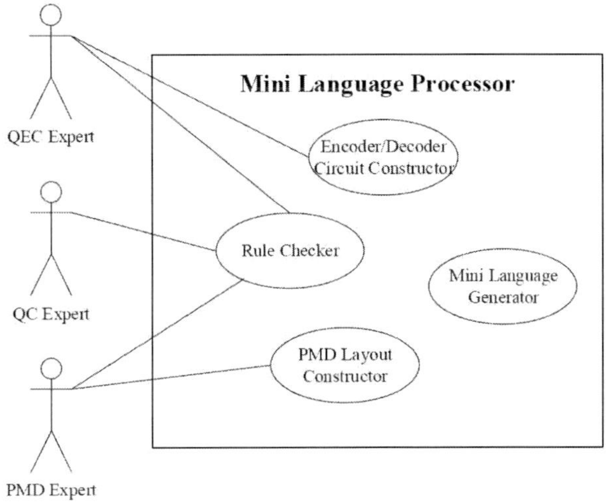

Figure 8. NQCS Mini language processor use-case diagram.

Table 3. The detailed description for the "Mini language" use-case

Use-case name	**Mini language**
Related requirements	Requirement 2 in Section 3
Goal in context	A new PMD, QC protocol, or QEC code is described by the PMD expert, QC expert, or QEC expert respectively
Preconditions	All XML and QASM files are available, and are specified correctly.
Successful end conditions	A new PMD, QC protocol, or QEC code is integrated into NQCS.
Fail end conditions	The PMD, QC protocol, or QECC specification is rejected.
Primary actors	PMD expert, QC expert, or QEC expert
Secondary actors	-
Trigger	The PMD expert, QC expert, or QEC expert asks the NQCS to process the new PMD specification, QC protocol, or QEC code, respectively.

achieving it. We anticipate that a totally new QEC will require changes to the NQCS toolbox that go beyond the mini-language processor and touch on other aspects of the toolbox including the front- and back-end tools in potentially significant ways.

For EFFBD of the NQCS Mini language processor refer to Figure 5.

4.4.2. Circuit Construction/Optimization

A *circuit construction/optimization* tool [16] (cf. function 2.4 in Figure 5, and expansion in Figure 9) is used in the Mini language processor. This tool finds optimized quantum circuit realizations for compound gates (i.e., gates that require more than one native instruction to be realized in the given PMD). The optimization objective may be the number of native instructions or the execution time for the compound gate operation. The tool relies on a *default - (O2I) mapping tool*, or *O2I mapper* for short, (function 2.4.1 in Figure 9) to compute and store in an O2I library the optimal realizations of each basic operation in terms of native instructions in the given PMD. Optimal realization means either the minimum number of native instructions or the shortest execution time. The native instruction set for this tool are: (i) One-qubit operations (R_x, R_y, R_z, X, Y, Z, S, T, H), and (ii) two-qubit operations (CNOT, CZ, Geometric, Controlled-Phase, Swap, iSwap). After mapping the circuit into native instructions that are supported by the PMD, some adjacent instructions may be redundant. Thus, the tool optimizes the circuit by using quantum identity rules (function 2.4.2 in Figure 9) until no more optimizations are possible. Figure 10 shows the construction of SWAP gate in various PMDs, obtained by the circuit construction/optimization tool [16].

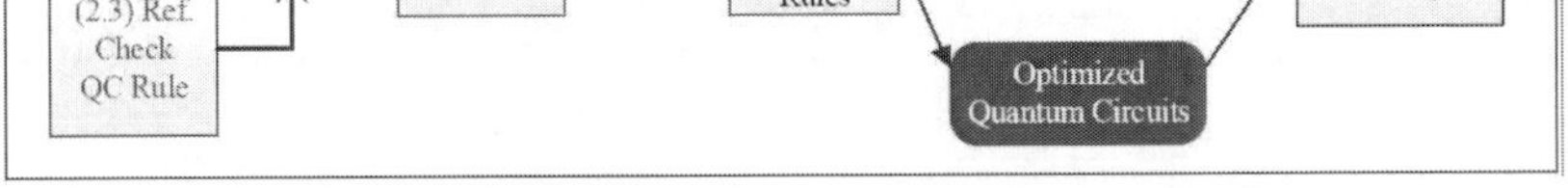

Figure 9. NQCS circuit constructor/optimizer.

4.5. Back-end

The back-end tool suite comprises of a - (QPD) tool, which is shown in Figure 11 with its use-case model in Figure 12. The inputs to the QPD are the

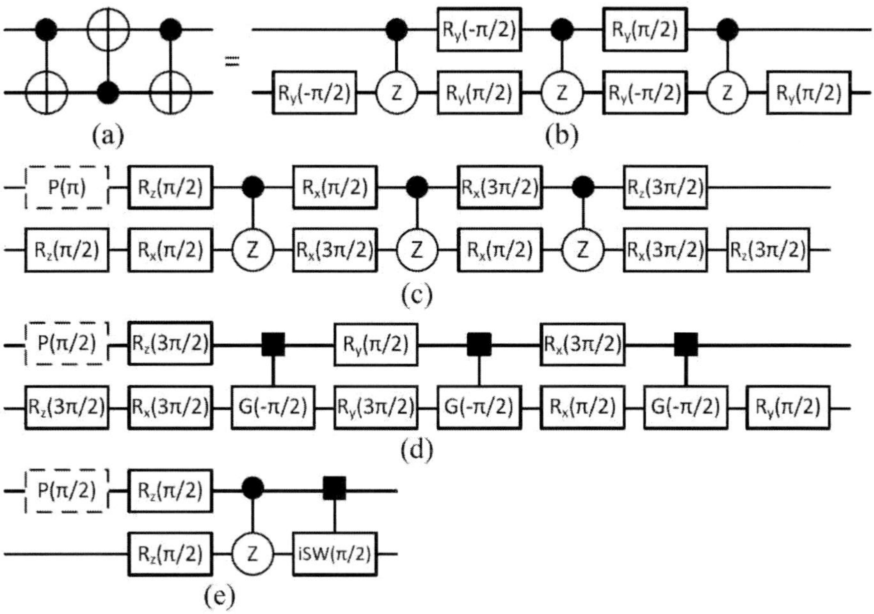

Figure 10. SWAP gate construction in (a) photonics, (b) neutral atom, (c) quantum dot, (d) trapped ion, and (e) superconducting PMDs [16].

HF-QASM representation, which is the output of the front-end tool suite, and tile design/tiled architecture along with gate models, which are outputs of the library designer. The HF-QASM representation is translated into a QIDG, which is subsequently used during scheduling, placement, and routing steps (see next paragraph). After scheduling, placement, and routing steps, a new HF-QASM is generated, which includes information about the operation locations (the placement coordinates of the tile in the quantum fabric where each operation takes place) as well as qubit movements (the steps that specify how logical qubits are routed between tiles). By substituting fault-tolerant quantum gates and move operations with respective MCL codes, this HF-QASM is translated into the circuit MCL code. Applying syndrome extraction to this circuit based on operation, movement, and stall error rates is also the responsibility of back-end when generating the final circuit MCL code.

Scheduling, placement, and routing problems for the back-end are defined in the same way as those for the library designer. However, the back-end deals

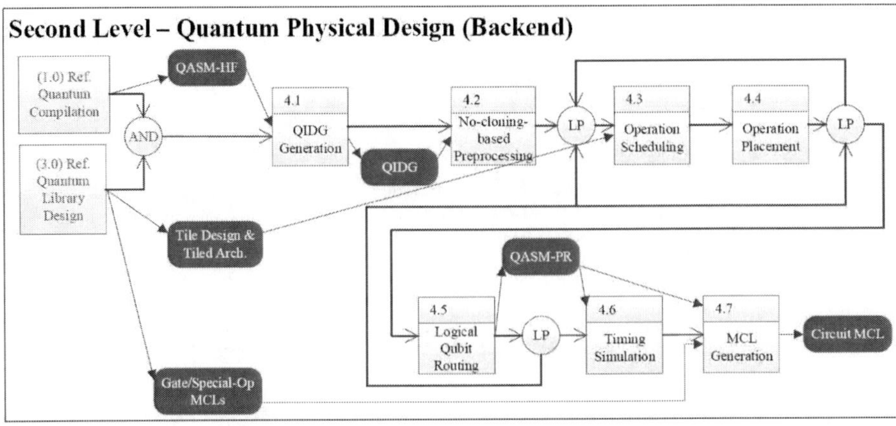

Figure 11. EFFBD of NQCS back-end.

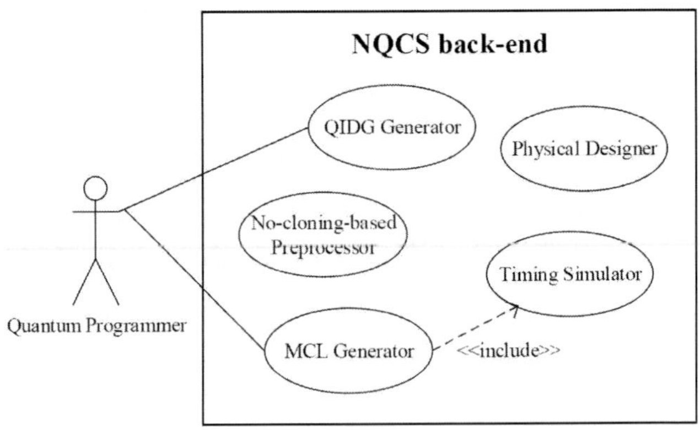

Figure 12. NQCS back-end use-case diagram.

with logical qubits and operations, as opposed to the library designer that deals with physical qubits and operations. The library designer receives the physical realization of a basic operation, schedules and places native instructions on the basic cells of the quantum fabric, and routes physical qubits between these cells in order to generate the MCL code for the given operation. On the other hand, the input to the back-end is a circuit potentially composed of a very large number of basic operations on logical qubits (i.e., the HF-QASM representation). The MCL generation of the circuit in the back-end tool proceeds by scheduling and placing these operations on the 2D tiled architecture, routing logical qubits between tiles, and replacing various operations and moves with respective MCL codes. Another key difference arises from the scale (size) of the input file (problem). While the number of native instructions in a fault-tolerant implementation of a basic operation can vary from hundreds to thousands of instructions, a HF-QASM representation may contain (tens of) millions of operations [19]. To handle this huge input size and generate a scalable back-end, partitioning techniques on QIDG and tiled-architecture are necessary. Using partitioning techniques, the number of operations per partition can be limited to a reasonable number for placement and routing algorithms. The required steps for MCL generation and timing simulation are summarized in Table 4 and Table 5.

4.5.1. Timing Simulation

After generating the MCL code by the back-end tool suite or in the case of constructing MCL code for each fault-tolerant quantum gate by the - (QUFD), scheduling, placement, and routing steps are visualized by a timing simulation tool. The main function of timing simulation is to extract timing information, i.e., starting execution time of each instruction, latency of routing qubits, and the delay of the quantum circuit. As a byproduct, timing simulation may work as a verifier which examines the correctness (and possibly the efficiency considering different metrics) of scheduling, placement, and routing solutions. Table 5 summarizes the description.

4.6. - (RC)

The RC function, as shown in Figure 13 with use-case diagram in Figure 14, has been automated in the NQCS toolbox. After compiling the Scaffold code, the HF-QASM is automatically analyzed to produce the number of basic operations, and the estimated parallelism and communication cost for routing the qubits (cf. function 5.1 in Figure 13). The error rate and delay information for basic operations are also automatically extracted from the machine description and protocol libraries (cf. function 5.2 in Figure 13). The number of gates and the time needed to perform quantum error correction that yields a desired error rate is expressed as a series of recursive functions in the automatic - (ECC) analysis tool (cf. function 5.3 in Figure 13). If the error threshold is met, the *ECC level/distance calculation* tool (cf. function 5.4 in Figure 13) computes the concatenation level (in case of concatenated codes), or the code distance (in

Table 4. The detailed description for the "Generate MCL" use-case

Use-case name	**Generate MCL**
Related requirements	Requirement 4 in Section 3
Goal in context	- (MCL) is generated.
Preconditions	The Scaffold program should be compiled with no error.
Successful end conditions	MCL for the quantum algorithm is generated.
Fail end conditions	An error message is generated.
Primary actors	Quantum Programmer
Secondary actors	-
Trigger	The quantum programmer asks the NQCS to generate the MCL code.
Included Cases	Simulate Timing
Main flow	Action
	(Step #1) For each QGS operation as well as special-ops an MCL is generated. (Step #2) [include::Simulate Timing] Timing of each MCL is verified by timing simulation. (Step #3) The HF-QASM representation is translated into MCL code based on a 2D tiled architecture and MCL codes from step 1. (Step #4) [include::Simulate Timing] Timing of the MCL code is verified by timing simulation.

case of surface codes) based on (i) the number of basic operations that comes from function 5.1, (ii) the gate error rates that come from function 5.2, and (iii) the functions that come from function 5.3. Finally, a *resource calculation* tool (cf. function 5.5 in Figure 13) combines these outputs, and reports the number of physical qubits, number of physical gate operations, and number of clock cycles required to execute the quantum algorithm. However, in the case that the error rate of a quantum gate (after applying the QC protocol) is higher than the error threshold of the QEC code, the given combination of PMD, QC, and QEC is not allowed (cf. function 5.6 in Figure 13). The required steps for PMD are

Table 5. The detailed description for the "Simulate Timing" use-case

Use-case name	**Simulate Timing**
Related requirements	Requirement 6 in Section 3
Goal in context	Latency of a PMD-based specification (i.e., MCL of a QGS operation, Special-op, or the circuit) is determined. Circuit will be visualized for possible actions from quantum programmer.
Preconditions	MCL code is generated successfully.
Successful end conditions	Timing information and other related information for the MCL code is displayed.
Fail end conditions	An error message is displayed.
Primary actors	Quantum Programmer
Secondary actors	-
Trigger	The quantum programmer asks the NQCS to run the timing simulator.

summarized in Table 6.

As discussed in [19] in detail, a suite of Octave scripts has been provided that automatically generates the resource estimates for a cross product of algorithms, PMDs, and error correction techniques. For each quantum algorithm, the number of logical gates of each type, the parallelization factor for these gates (how many gates of each type can be safely scheduled in parallel), and the number of logical qubits are needed. Another required input is information about each combination of PMD and control protocol. The required information is the time of each gate type, the error of the worst gate, and error rate per unit of time. For more information, please refer to [19].

5. Interface Specification

N2 diagram (or N × N interaction matrix) [12] is used in this section to identify interactions or interfaces between major functions of the NQCS toolbox from a system perspective. This diagram is typically used to develop system interfaces. The N2 diagram can be taken down into successively lower levels to the component functional levels. In addition to defining the interfaces, the

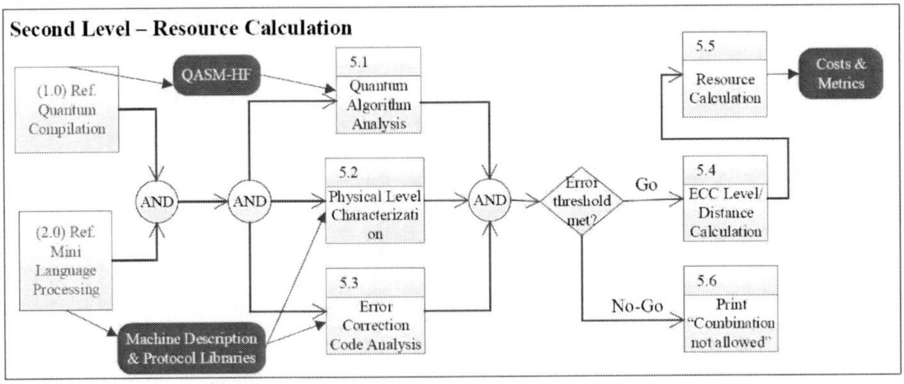

Figure 13. EFFBD of NQCS - (RC).

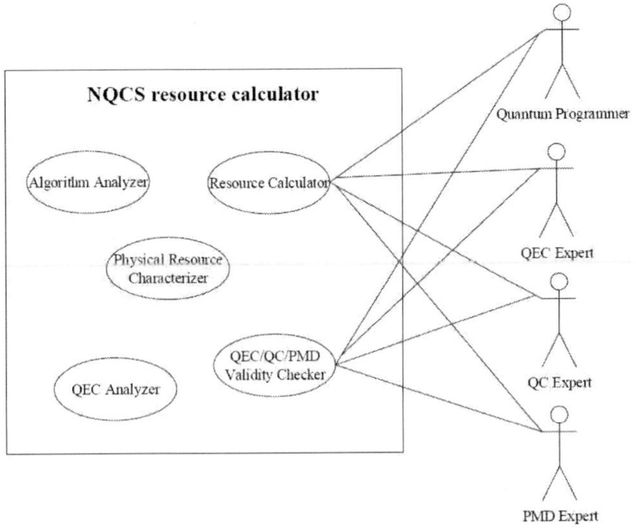

Figure 14. NQCS - (RC) use-case diagram.

N2 diagram also pinpoints areas where conflicts could arise in interfaces, and highlights input and output dependency assumptions and requirements.

The top-level N2 diagram for the NQCS toolbox is depicted in Figure 15. In this diagram, various NQCS functions are placed on the diagonal. The remainder of the squares in the N x N matrix represents the interface inputs and outputs. A blue geometric shape at the intersection of a row and a column con-

Table 6. The detailed description for the "Calculate Physical Resources" use-case

Use-case name	**Calculate Physical Resources**
Related requirements	Requirement 5 in Section 3
Goal in context	Physical resources for a given HF-QASM, PMD, QC, and QEC are calculated.
Preconditions	HF-QASM is generated, and PMD, QC protocol and QEC code are described.
Successful end conditions	Physical resource calculation is displayed.
Fail end conditions	An error message ("combination not allowed") is displayed.
Primary actors	Quantum Programmer
Secondary actors	PMD Expert, QC Expert, QEC Expert
Trigger	The Quantum Programmer asks the NQCS to launch the resource calculator.
Main flow	Action
	(Step #1) Scaffold program is compiled into HF-QASM in order to find the number of basic operations, and the estimated parallelism and communication cost. (Step #2) Error rate and delay information of basic operations are extracted from PMD and QC protocols. (Step #3) Error threshold is extracted from QEC code. (Step #4) If the error threshold is met, concatenation level or the code distance is computed, and then the number of physical qubits, number of physical gate operations, and number of clock cycles required to execute the quantum algorithm are reported. (Step #5) Otherwise, the given combination of PMD, QC, and QEC is not allowed.

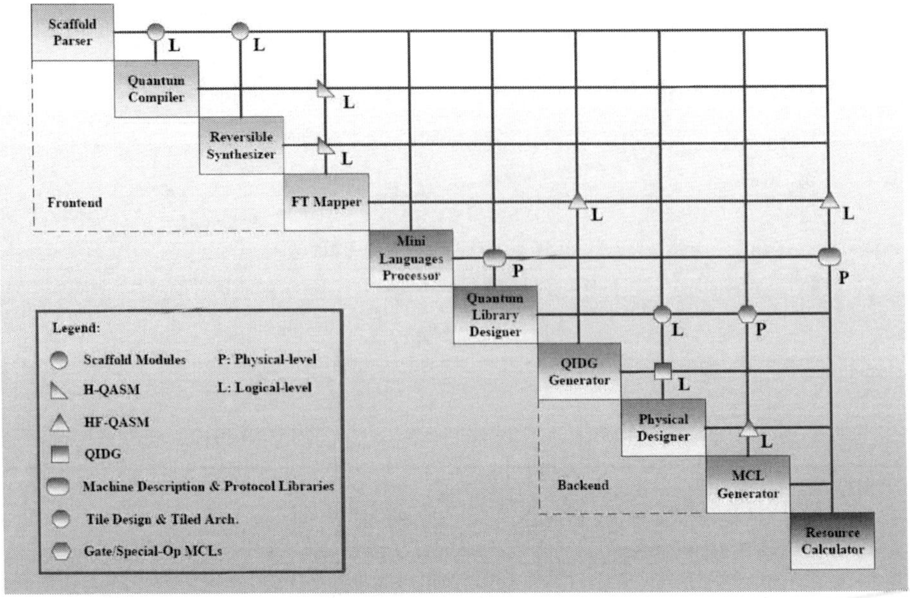

Figure 15. N2 Diagram of the NQCS toolbox.

```
#define N 4

module cat ( qbit *bit, const int n ) {
    H( bit[0] );
    for ( int i=1; i < n; i++ ) {
        CNOT( bit[i], bit[i-1] );
    }
}

int main () {
    qbit bits[N];
    cat( bits, N );
    return 0;
}
```

Figure 16. The Scaffold code for the cat state preparation circuit.

tains a description of the interface between the two functions represented on that row and that column. A letter “L” (“P”) at the intersection of a row and a column indicates an interface that acts on logical (physical) qubits/gates between the two functions. For example, the - (FT) mapper has a logical interface, which is called HF-QASM, with the resource calculation tool. Where no shape appears, there is no interface between the respective functions.

The rest of this section explains the main interfaces of the NQCS toolbox in more details. The interface examples are for demonstration purposes only.

1. **Scaffold code:** This language is developed as a - (QPL). The details of Scaffold are presented in a separate document [10]. As a working example, we can consider the Scaffold code shown in Figure 16, which is the cat state preparation circuit.

2. **H-QASM:** This representation is the output of the reversible logic synthesis tool as well as the quantum compiler, and the input to the fault-tolerant mapper. The main difference between H-QASM and the standard QASM [11] is the ability to use hierarchical design, which is important for complex circuits. More precisely, H-QASM adds the following features to the standard QASM representation.

   - An entity called "module" which defines a quantum sub-circuit. The sub-circuit then can be reused several times.
   - Any module or gate can be repeated several times by indicating the "repeat" keyword before the respective module/gate.

   Although modular design simplifies the design process, the aforementioned features are vital to maintain the code size reasonable. Figure 17(b) shows the H-QASM grammar.

3. **HF-QASM representation**: This file is the output of the fault-tolerant mapper. HF-QASM inherits all features of H-QASM; however, it limits the gate sets to only fault-tolerant gates. The grammar of HF-QASM is shown in Figure 17(c) [9]. HF-QASM allows the fault-tolerant mapper to define new modules (or gates) such as Toffoli gates. The resultant file should provide the implementation of all gates using only fault-tolerant gates (i.e., QGS). The HF-QASM code for the cat state preparation circuit generated by the Scaffold compiler is shown in Figure 18.

4. **QIDG:** After processing the HF-QASM, a *quantum instruction dependency graph* (QIDG) is generated. Instruction dependencies avoid -

(WAR), - (WAW), and - (RAW) hazards. In addition, the quantum no-cloning theorem adds another dependency to this graph that avoids multiple reads of a qubit at the same time. This graph is used in scheduling, placement and routing steps of the quantum physical designer.

5. **Machine Description and Protocol Libraries**: Mini languages specify the target physical machine description, applied quantum control scheme and quantum error correction code in the quantum circuit. By fast processing of this information, it is possible to produce basic estimates about performance metrics of each gate in the quantum circuit. These estimates contain the number of native instructions needed to implement each gate and rough estimates of the total latency and error probability of their physical implementation. These estimates can be used to derive the baseline resource estimate of the quantum circuit.

6. **Tile Design and Tiled Architecture**: Quantum library designer examines different tile sizes and tile architectures to find the best tile architecture in terms of the quantum circuit performance metrics. The tile architecture specifies the size of the tiles, location of the tiles and topology of the whole fabric. It is used in the back-end for scheduling, placement and routing of logical qubits. Gate models specify the performance metrics, i.e., latency and error probability, of the required gates in the quantum circuit.

7. **MCL code**: Machine control language code specifies the lowest level control commands for the target quantum circuit.

6. Verification

Verification of a product shows proof of compliance with requirements — that the product can meet each shall statement as proven through performance of a test, analysis, inspection, or demonstration. More precisely, verification is answering to the following question: Does the toolbox perform what it claims to do correctly?[4]. To manage the complexity and ensure scalability, we have opted

[4]On the other hand, validation of a product shows that the product accomplishes the *intended purpose* in the intended environment — that the product meets the expectations of the customer and other stakeholders as shown through performance of a test, analysis, inspection, or demonstration. Validation is answering to the following question: Is the designer claiming that the

to break the verification process of the NQCS toolbox into smaller verification processes which includes "verification of algorithm to QPL", "verification of toolbox design requirements", and "verification of QPL to MCL". The difference between the first and third tasks in verification arises from the potential source of errors in Scaffold code and the MCL code. More precisely, considering the manual process of coding a quantum algorithm, the main problem in Scaffold code can arise from software programming mistakes. In contrast, considering a given Scaffold code, errors in the MCL code are the result of the mistakes in the toolbox design and tool suite. In addition to these mistakes, the MCL code should be verified to make sure that the error threshold of the applied QEC scheme is greater than the error probability of the PMD instructions considering the QC scheme. The EFFBD of NQCS toolbox verification process is shown in Figure 19.

Table 7. The detailed description for the "Verify Implemented Algorithm" use-case

Use-case name	Verify Implemented Algorithm
Related requirements	Requirement 7 in Section 3
Goal in context	To make sure that the Scaffold program satisfies all the requirements of the quantum algorithm.
Preconditions	The Scaffold program should be given.
Successful end conditions	The Scaffold program has a reasonable chance of correctly implementing the quantum algorithm.
Fail end conditions	The Scaffold program is not verified at all.
Primary actors	Quantum Programmer
Secondary actors	-
Trigger	The quantum programmer asks the NQCS to verify the Scaffold program.
Main flow	Action
	Debugging.

toolbox does the right things?

6.1. Verification of Algorithm to QPL

Quantum algorithm to QPL Verification (Table 7) requires that we provide a proof that the QPL code correctly implements the desired algorithm. This is an extremely challenging task that does not lend itself to an exact solution. In general, the problem of automatic validation of a software program is intractable even in classical computing. Similarly, the problem of automatic validation of a Scaffold code versus a given quantum algorithm is intractable (and in fact undecidable). A viable option is to run different benchmark circuits and compare the outputs with expected (golden) results. Another option is to ask the QPL programmer/user to provide a proof of the correctness for his code using formal methods (although this may be a huge demand imposed on the user).

Verification for larger quantum circuits will be done through inspection. When possible, arguments will be constructed to show that small-input verification is indicative of large-input cases. At the QPL level, we expect standard compiler techniques to emit code that maintains correctness at large scale. Such techniques have been shown to be behaviorally correct over the long history of compiler usage for large-scale classical programs and inputs. Since each algorithm is only one or two thousand lines of Scaffold code, we expect verification by inspection to be a reasonable approach.

In this way, we reduce the number of casual errors introduced by software programmers.

6.2. Verification of Toolbox Design Requirements

Each of the toolbox requirements (cf. Section 3) should be evaluated by one specific testing strategy to make sure that each requirement is properly addressed. Excluding the topics discussed in the other sections leads to the following items:

1. The NQCS toolbox shall allow the quantum programmer to write the quantum algorithm in Scaffold providing quantum data types, quantum operations, classical operations, and appropriate control flow structures. This can be evaluated by "demonstration" — by expressing several quantum algorithms by Scaffold.

2. The NQCS toolbox shall enable (1) PMD expert to describe a new PMD architecture, along with error rates and latencies for various quantum instructions, (2) QEC expert to describe a new quantum error correction

code, and (3) QC expert to describe a new quantum control protocol. These requirements can be evaluated by "demonstration" — by handling different QCs, QECs, and PMDs.

3. The NQCS toolbox shall provide an efficient quantum compiler for the quantum programmer. This can be checked by comparison of compiler outputs for several well-defined problem sizes with the expected outputs.

4. The NQCS toolbox shall provide a set of software utilities such as a timing simulator as the NQCS quantum programming environment. This requirement can be verified by "demonstration" — (1) by manual output comparison for several specific small cases, and (2) by automatic output comparison for many large test vectors with known results.

6.3. Verification of QPL to MCL

The goal of this step is to ensure that the generated MCL code is a valid realization of the Scaffold code on the target PMD. We plan to divide this verification task into three subtasks: (i) verification of the HF-QASM against the Scaffold code, (ii) verification of the FT gate library, and (iii) verification of the MCL code against HF-QASM.

- Different parts of the Scaffold code may follow different synthesis/compilation paths to be transformed into the HF-QASM. This motivates us to take different verification approaches for these parts.

  - Reversible circuits can be reduced to conventional AND-OR-NOT circuits for purposes of verification. As a result, for the classical modules and overall control flow, classical verification algorithms and tools similar to Mentor Graphics QuestaSim[5] can be used. Additionally, equivalence checking algorithms for reversible circuits can be used [20]. Equivalence checking is a common term in the published papers and is used to check two circuits (or a circuit vs. a truth-table).
  - For standard quantum modules such as quantum Fourier transform or realization of arbitrary rotations using the Solovay-Kitaev algorithm [14], it is possible to do the verification by construction because all of these quantum modules will be replaced by HF-QASM

[5] http://www.mentor.com/products/fv/questa/

blocks by using well-known (textbook) algorithms that have analytical proof of correctness. As an alternative approach, recent verification algorithms [21], [22] for quantum logic can also be used.[6]

- Generating the FT gate library needs both verification and error simulation steps. Similar to the module responsible for generating FT gate library, related verification and error simulation steps can be performed once, and cached to be reused later on by the toolbox as needed. It is, of course, impossible to simulate arbitrary quantum circuits, or there would be no need to build quantum computers; but large pieces of FT code can be simulated using efficient techniques.

- To verify the generated MCL code for each FT gate in the library against the transformation matrix of the FT gate, we will exploit the Gottesman-Knill theorem [6][7]. More precisely, we will use [23]. This simulation can be used for an adequate number of input (stabilizer) states to guarantee correct behavior with reasonably high probability. In a similar way, the classical modules (e.g., oracle functions) are essentially reversible classical circuits; they can be checked for random computational basis states to give probabilistic verification of their correctness.

- In addition to this quantum simulation tool, a Monte Carlo simulation framework can be developed to check if the applied QEC satisfies the maximum error probability and to determine the success rate of the gate. This can be done in a simplistic manner merely by generating random errors and estimating overall failure rates.

[6]PP (Probabilistic Polynomial-Time) is the class of decision problems solvable by an NP (Nondeterministic Polynomial-Time) machine which gives the correct answer (i.e., 'Yes' or 'No') with probability $> \frac{1}{2}$. P^{PP} (P with PP oracle) includes decision problems solvable in polynomial time with the help of an oracle for solving problems from PP. Quantum circuit simulation belongs to the complexity class P^{PP} [22].

[7]According to the Gottesman-Knill theorem [6], quantum circuits exclusively consisting of the following components can be efficiently simulated on a classical computer in polynomial time:

- A state preparation N-circuit with initial value $|000...0\rangle$ — qubit preparation in the computational basis,
- Quantum gates from the Clifford group (Hadamard, Phase, CNOT, and Pauli gates),
- Measurements in the computational basis

- Error probability or success rate of the gates can be used in error analysis of the final MCL code to check the total success rate of the quantum circuit. Verification of the MCL code against HF-QASM is not the target of our toolbox because we expect that the number of FT gates to be too large to be simulated using classical computers.

7. Performance Measures

Performance measures are the non-functional requirements of the NQCS toolbox. They are used to assess (i) the quality of results and (ii) the scalability of the NQCS toolbox.

7.1. Quality of Results

We use the following performance measures to evaluate the outputs of the toolbox.

- Total number of physical qubits: The total number of physical qubits required to implement an algorithm.
- Circuit depth: This measure reflects the total runtime of the algorithm considering the degree of parallelization in resulting circuits.
- Number of ancillae: Along with the number of main qubits, the number of ancillae required at each step is an important parameter for circuit evaluation.
- Runtime/area ratio of the ideal circuit to the encoded circuit: The ratio of runtime/area metric for the ideal circuit to that for the QEC-enabled circuit reflects the efficiency of the QEC code.
- Area-Delay-to-Correct-Result: This metric is introduced in [24] and is defined as follows:

$$ADCR = Area \times \frac{Latency}{P_{success}}, \tag{1}$$

  where $Area$ is the area of the bounding box in the PMD in which the quantum algorithm is mapped to, $Latency$ is the delay of execution of the quantum algorithm, which is calculated by the timing simulator, and $P_{success}$ is the probability of getting the correct (error-free) answer.

name → **[a-z,A-Z]+[a-z,A-Z,0-9]***
num → **[1-9][1-9]***
whitespace → **\n** — **\r** — **\t** —
comment → **#** .* (**\n** — **<EOF >**)

(a)

start → (module)* main
main → **module main {** body **}**
module → **module** name **(** param_list **) {** body **}**
param_list → param (,param)*
param → **qubit** (*)? name
body → (def;)+ (gate;)+
def → **qubit** ([num])? name
gate → (one_qubit_gate — multi_qubit_gate — call)
one_qubit_gate → (**H**—**X**—**Y**—**Z**—**S**—**S**†—**T**—**T**†—**Prep0**—**MeasX**—**MeasY**—**MeasZ**) (arg)
multi_qubit_gate → (**Toffoli** (ctrlArray, arg)) — (**Fredkin** ((ctrlArray,)? arg, arg))
call → name (call_list)
call_list → arg (, arg)*
arg → name — name[num]
ctrlArray → arg (, arg)*

(b)

start → (module)* main
main → **module main {** body **}**
module → **module** name **(** param_list **) {** body **}**
param_list → param (,param)*
param → **qubit** (*)? name
body → (def;)+ (gate;)+
def → **qubit** ([num])? name
gate → (one_qubit_gate — two_qubit_gate — call)
one_qubit_gate → (**H**—**X**—**Y**—**Z**—**S**—**S**†—**T**—**T**†—**Prep0**—**MeasX**—**MeasY**—**MeasZ**) (arg)
two_qubit_gate → **CNOT** (arg, arg)
call → name (call_list)
call_list → arg (, arg)*
arg → name — name[num]

(c)

Figure 17. (a) Definition of tokens. (b) H-QASM grammar. (c) HF-QASM grammar [9]. Regular expression is used to simplify the grammar specification. Note that the star character (*) used for the *param* variable is a *terminal*. It determines whether a parameter is an array or not (similar to the C language).

Table 8. Comparison of QASM and HF-QASM file sizes for different quantum benchmarks. *module count* denotes the number of modules used in the benchmark [9].

Benchmark	Ref.	Module Count	Problem Size	File Size		Size Ratio
				QASM	HF-QASM	
Grover's Algorithm	[2]	6	n=100†	548KB	52KB	11
		6	n=300†	3.2MB	160KB	20
		6	n=500†	7.0MB	268KB	27
Binary Welded Tree	[25]	3	$n = 43, 3 \leq s \leq 19$‡	5.52MB – 34.9MB	84KB – 88KB	67 – 406
Ground State Estimation	[3]	170	M=6§	481MB	200KB	2,462
Triangle Finding Problem	[4]	332	n=5*	63GB	504KB	129,418
		10,202	n=10*	Failed¶	28MB	N/A¶

† A database of 2^n elements is being searched.

‡ n is the height of the tree and s is a time parameter within which the solution is found.

§ M is the molecular weight of a molecule.

* n is the number of nodes in the graph.

¶ The QASM file size exceeded 75GB and Scaffold compiler was crashed while generating the output file.

```
module cat_4 ( qbit* bit ) {
    H ( bit[0] );
    CNOT ( bit[1] , bit[0] );
    CNOT ( bit[2] , bit[1] );
    CNOT ( bit[3] , bit[2] );
}

module main ( ) {
    qbit bits[4];
    cat_4 ( bits );
}
```

Figure 18. The HF-QASM code for the cat state preparation circuit.

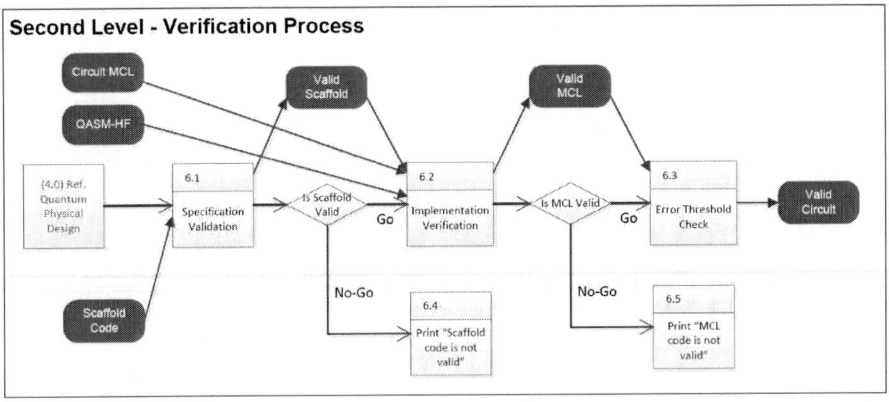

Figure 19. EFFBD of NQCS verification.

In addition to the aforesaid metrics, there are other metrics that can be considered during different steps in order to make sure that our circuits meet the final constraints. Among these metrics, geometric constraints (e.g., nearest neighbor interaction [17]), and gate complexity (e.g., the average number of control qubits of each gate) are adopted in our toolbox. Furthermore, to be amenable to automation, the toolbox can take into account the role of the researcher/user in various parts. One possibility is to provide the user access to manipulate the trade-offs between resources. As an example, our back-end tool allows the user to explore the circuit latency vs. ancilla count trade-off [26]. More specifically, physical ancilla qubits are precious resources in quantum computers. However, increasing the total ancilla budget may lead to a lower circuit latency and hence improved quality of results.

7.2. Scalability

The toolbox should be able to handle sufficiently large problem sizes. To better appreciate the scalability challenge, we mention a few notable results.

- For a reversible circuit with n lines, where its optimal gate-count realization, needs h gates from a library $\mathcal{L}$, an enumerative method may branch h ways on each $\mathcal{L}$-gate. For example, assume that only multiple-control Toffoli gates exist in the library. For this simplified case, an exhaustive method examines $(n \times 2^{n-1})^h$ gates[8] to find an optimal circuit. Currently, the sharpest upper bound on the number of gates for a reversible circuit is $8.5n2^n + o(2^n)$ [27]. Therefore, a given circuit may need an exponential number of gates.

- The method in [28] uses a search-based method [29] to synthesize a given function. Therefore, it may not be able to find a result. Accordingly, the NQCS toolbox uses [13] which adopts decision diagrams to handle large functions. The early synthesis results on an Intel 3.1 GHz CPU with 4 GB of memory with a runtime limit of 600 seconds reveal interesting results [13]. We can increase the runtime limit to several days and improve processing power in order to enhance the quality of results and also to handle large-scale quantum circuits.

- The search space for quantum-logic synthesis is not finite, and circuits implementing generic unitary matrices require $\Omega(4^n)$ gates [30]. We use the Solovay-Kitaev theorem for compiling an arbitrary single-qubit gate into a sequence of gates from a fixed and finite set. The algorithm runs in $O(\log^{2.71}(1/\epsilon))$ time.[9] Considering an exponentially number of gates leads to a runtime/memory limit on the classical computer assigned to

[8]There are $\binom{n}{1}$ possible NOT gates and $\binom{n}{2}$ possible CNOT gates in which one of its two inputs can be the target output. Hence, the total number of $2\times\binom{n}{2}$ CNOT gates can be obtained. For a $(k{+}1)$-bit gate, $k \in (2, 3, \cdots, n-1)$, there are $\binom{n-1}{k}$ possible gates when the target can be the i-th ($i \in [1, n]$) bit. Considering all possible bits as the target leads to $n \times \binom{n-1}{k}$ $(k{+}1)$-bit gates. Therefore, the total number of gates is $\binom{n}{1} + 2 \times \binom{n}{2} + n \times (\sum_{i\in(2\cdots n-1)} \binom{n-1}{i}) = n \times 2^{n-1}$.

[9]It produces as output a sequence of $O(\log^{3.97}(1/\epsilon))$ quantum gates which is guaranteed to approximate the desired quantum gate to an accuracy within $\epsilon > 0$.

run the algorithms. Therefore, the processing power of the underlying machine will limit the toolbox.

- Quantum circuit simulation belongs to the complexity class P^{PP}. Hence, handling an arbitrary-size circuit is out of reach in general. For reversible circuits, SAT-based techniques lead to good results [20]. While the Boolean satisfiability problem is NP-complete, the state-of-the-art SAT solvers solve many practical problem instances, i.e., a SAT instance can consist of hundreds of thousands of variables, millions of clauses, and tens of millions of literals. Hence, we expect reasonable results on a normal classical computer is obtainable. For quantum circuit verification, state-of-the-art techniques [22] use improved decision diagrams and can handle middle-size problems fast. However, automatic verification of large circuits is limited by constraints imposed by NP-complete and P^{PP} problems.

A key design strategy that significantly enhances the scalability of the NQCS toolbox is to take advantage of the hierarchical design of quantum circuits. More specifically, quantum circuits can be partitioned into repetitively-used quantum modules. This means that mapping one instance of these modules is sufficient. Accordingly, HF-QASM has been chosen as the main input to the back-end tool. Table 8 compares sizes of compiled codes in QASM and HF-QASM formats for various quantum benchmarks. As can be seen, for complex algorithms such as *Triangle Finding Problem*, QASM format is completely inefficient because it requires more than 60GB of disk space. On the other hand, HF-QASM can be successfully adopted. For more information, please refer to [9].

8. Further Reading

For further background on quantum information and computation, interested readers are encouraged to consult the main textbooks including the book by Nielsen and Chuang [6]. For quantum error correction techniques please refer to the book by Lidar and Brun [31]. For an introduction of reversible circuits and related synthesis and optimization methods refer to the recent survey by Saeedi and Markov [29]. For simulation and verification of quantum circuits please consult [22].

List of Acronyms

BAA - Broad Agency Announcement

BRE - Baseline Resource Estimation

CTQG - Classical code To Quantum Gate

CAD - Computer Aided Design

CULB - Computing Universal Logic Block

ECC - Error Correction Code

EFFBD - Enhanced Functional Flow Block Diagram

FT - Fault-Tolerant

GFI - Government Furnished Information

IARPA - Intelligence Advanced Research Projects Activity

ISA - Instruction Set Architecture

MCL - Machine Control Language

MCT - Multiple-Control Toffoli

MULB - Memory Universal Logic Block

NQCS - Next-generation Quantum Computing Systems

O2I - Operation-to-Instruction

PMD - Physical Machine Description

QASM - Quantum Assembly Language

H-QASM - Hierarchical Quantum Assembly Language

HF-QASM - Hierarchical Fault-tolerant Quantum Assembly Language

QC - Quantum Control

QCS - Quantum Computing Systems

QEC - Quantum Error Correction

QGS - Quantum Gate Set

QIDG - Quantum Instruction Dependency Graph

QPD - Quantum Physical Designer

QPL - Quantum Programming Language

QUFD - Quantum Universal logic block Factory Designer

RAW - Read After Write

RC - Resource Calculation

RMDDS - Reed-Muller Decision Diagram Synthesis

SAT - Satisfiability

TCT - Technical Coordination Team

ULB - Universal Logic Block

WAR - Write After Read

WAW - Write After Write

XML - Extensible Markup Language

Acknowledgments

The authors would like to thank Prof. Todd Brun, Dr. Hadi Goudarzi, Ali Java-diAbhari, Dr. Chia-Chun Lin, Dr. Mehdi Saeedi, and Dr. Martin Suchura for their helpful discussions and inputs.

This research was supported in part by the Intelligence Advanced Research Projects Activity (IARPA) via Department of Interior National Business Center contract number D11PC20165.

References

[1] P. W. Shor, "Polynomial-time algorithms for prime factorization and discrete logarithms on a quantum computer," *SIAM Journal on Computing*, vol. 26, no. 5, pp. 1484–1509, Oct. 1997.

[2] L. K. Grover, "A fast quantum mechanical algorithm for database search," in *Proceedings of the Theory of Computing*, 1996, pp. 212–219.

[3] J. D. Whitfield, J. Biamonte, and A. Aspuru-Guzik, "Simulation of electronic structure hamiltonians using quantum computers," *Molecular Physics*, vol. 109, no. 5, pp. 735–750, 2011.

[4] F. Magniez, M. Santha, and M. Szegedy, "Quantum algorithms for the triangle problem," *SIAM Journal on Computing*, vol. 37, no. 2, pp. 413–424, 2007.

[5] T. D. Ladd, F. Jelezko, R. Laflamme, Y. Nakamura, C. Monroe, and J. L. O'Brien, "Quantum computers," *Nature*, vol. 464, no. 7285, pp. 45–53, 2010.

[6] M. A. Nielsen and I. L. Chuang, *Quantum Computation and Quantum Information*. Cambridge University Press, 2000.

[7] A. JavadiAbhari, S. Patil, D. Kudrow, J. Heckey, A. Lvov, F. Chong, and M. Martonosi, "ScaffCC: A Framework for Compilation and Analysis of Quantum Computing Programs," in *International Conference on Computing Frontiers*, May 2014.

[8] H. Goudarzi, M. J. Dousti, A. Shafaei, and M. Pedram, "Design of a universal logic block for fault tolerant realization of any logic operation in an ion-trap quantum circuit," *Quantum Information Processing*, vol. 13, no. 5, pp. 1267–1299, Jan. 2014.

[9] M. J. Dousti, A. Shafaei, and M. Pedram, "*Squash 2: A Hierarchical Scalable Quantum Mapper Considering Ancilla Sharing*," submitted.

[10] A. Javadi Abhari, A. Faruque, M. J. Dousti, L. Svec, O. Catu, A. Chakrabati, C.-F. Chiang, S. Vanderwilt, J. Black, and F. Chong, "Scaffold: Quantum Programming Language," Department of Computer Science, Princeton University, Tech. Rep. TR-934-12, June 2012.

[11] K. M. Svore, A. V. Aho, A. W. Cross, I. Chuang, and I. L. Markov, "A layered software architecture for quantum computing design tools," *Computer*, vol. 39, no. 1, pp. 74–83, Jan. 2006.

[12] N. E. Rainwater and S. J. Kapurch, "NASA systems engineering handbook," *National Aeronautics Space Administration, Washington DC, NASA SP-2007-6105 Rev 1*, Dec. 2007.

[13] C. Lin and N. K. Jha, "RMDDS: Reed-Muller decision diagram synthesis of reversible logic circuits," *Journal on Emerging Technologies in Computing Systems*, 2013.

[14] C. M. Dawson and M. A. Nielsen, "The Solovay-Kitaev algorithm," *Quantum Information & Computation*, vol. 6, no. 1, pp. 81–95, Jan. 2006.

[15] D. Kudrow, K. Bier, Z. Deng, D. Franklin, Y. Tomita, K. R. Brown, and F. T. Chong, "Quantum rotations: a case study in static and dynamic machine-code generation for quantum computers," in *Proceedings of the International Symposium on Computer Architecture*, 2013, pp. 166–176.

[16] C. Lin, A. Chakrabarti, and N. K. Jha, "Optimized quantum gate library for various physical machine descriptions," *IEEE Transactions on VLSI Systems*, 2013.

[17] M. Saeedi, R. Wille, and R. Drechsler, "Synthesis of quantum circuits for linear nearest neighbor architectures," *Quantum Information Processing*, vol. 10, no. 3, pp. 355–377, Jun. 2011.

[18] E. Chi, S. A. Lyon, and M. Martonosi, "Tailoring quantum architectures to implementation style: a quantum computer for mobile and persistent qubits," in *Proceedings of the International Symposium on Computer Architecture*, 2007, pp. 198–209.

[19] M. Suchara, A. Faruque, C.-Y. Lai, G. Paz, F. Chong, and J. Kubiatowicz, "QuRE: The Quantum Resource Estimator Toolbox," in *International Conference on Computer Design*, October 2013.

[20] R. Wille, D. Große, D. M. Miller, and R. Drechsler, "Equivalence checking of reversible circuits," *International Symposium on Multiple-Valued Logic*, pp. 324–330, 2009.

[21] S.-A. Wang, C.-Y. Lu, I.-M. Tsai, and S.-Y. Kuo, "An XQDD-based verification method for quantum circuits," *IEICE Transactions on Fundamentals of Electronics, Communications and Computer Sciences*, vol. E91-A, no. 2, pp. 584–594, Feb. 2008.

[22] G. F. Viamontes, I. L. Markov, and J. P. Hayes, *Quantum Circuit Simulation*. Springer, 2009.

[23] S. Aaronson and D. Gottesman, "Improved simulation of stabilizer circuits," *Physical Review A*, vol. 70, p. 052328, Nov. 2004.

[24] M. G. Whitney, N. Isailovic, Y. Patel, and J. Kubiatowicz, "A fault tolerant, area efficient architecture for Shor's factoring algorithm," in *Proceedings of the International Symposium on Computer Architecture*, 2009, pp. 383–394.

[25] A. M. Childs, R. Cleve, E. Deotto, E. Farhi, S. Gutmann, and D. A. Spielman, "Exponential algorithmic speedup by a quantum walk," in *Proceedings of the Theory of Computing*, 2003.

[26] M. J. Dousti, A. Shafaei, and M. Pedram, "Squash: A Scalable Quantum Mapper Considering Ancilla Sharing," in *Proceedings of Great Lakes Symposium on VLSI*, May 2014, pp. 117–122.

[27] M. Saeedi, M. S. Zamani, M. Sedighi, and Z. Sasanian, "Reversible circuit synthesis using a cycle-based approach," *Journal on Emerging Technologies in Computing Systems*, vol. 6, no. 4, pp. 13:1–13:26, Dec 2010.

[28] P. Gupta, A. Agrawal, and N. Jha, "An algorithm for synthesis of reversible logic circuits," *IEEE Transactions on Computer-Aided Design of Integrated Circuits and Systems*, vol. 25, no. 11, pp. 2317 –2330, Nov. 2006.

[29] M. Saeedi and I. L. Markov, "Synthesis and optimization of reversible circuits — a survey," *ACM Computing Surveys*, vol. 45, no. 2, pp. 21:1–21:34, Mar. 2013.

[30] V. V. Shende, I. L. Markov, and S. S. Bullock, "Minimal universal two-qubit quantum circuits," *Physical Review A*, vol. 69, pp. 062 321:1–062 321:8, 2004.

[31] D. A. Lidar and T. A. Brun, *Quantum Error Correction*. Cambridge University Press, 2013.

In: New Developments in Quantum Optics ... ISBN: 978-1-63483-394-3
Editor: Nicole Stewart, pp. 51-97

Chapter 2

ENTANGLEMENT AND COOPERATIVE EFFECTS BETWEEN THE MODE COMPONENTS OF RAMAN PROCESS IN CAVITY AND THEIR ANALOGY WITH ATOMIC COLLECTIVE EFFECTS*

Nicolae A. Enaki†
Quantum Optics and Kinetic Processes Lab
Institute of Applied Physics, Academy of Sciences of Moldova,
Chisinau, Republic of Moldova

It is proposed the cooperative excitation and absorption of three cavity modes which corresponds to pump, Stokes and anti-Stokes photons stimulated by the atomic inversion. The three modes collective Roman emission is defined introducing the cooperative description between photon of cavity photons. In the case, when the scattering rates in the Stokes and anti-Stokes modes coincide, the SU(2) and SU(1,1) symmetries are applied for a simple description of these cooperative processes. The possibilities to realize this effect in the free space is proposed, replacing the cavity modes with dipole active excited atoms in tow-quantum interaction with dipole-forbidden transitions of D atom. The statistical properties and detection method are proposed using the information entropy and atomic correlation functions.

---

*The Author thanks for invitation.
†E-mail address: enakinicolae@yahoo.com; Website: http://www.qol.asm.md.

Abstract

The research was focused on studying the effects of cooperation between the Stokes and anti Stkes modes in Raman scattering processes. The phenomenon of quantum collectivization of photons scattered between three modes: pump, Stokes and anti-Stokes modes occur due to energy transfer between these waves according to the corpuscular theory of light. It was established the quantum properties of field in the resonator containing all three field modes pump Stokes and anti-Sokes. According to the conception of minimization of the energy of interaction between these three waves, we indicate the possibility of quantum coherence between Stokes and anti-Stokes modes, which possesses the new properties and possibilities for quantum processing of information introduced in the amplitude (or phase) of a bimodal field. The new coherent model shows us the existence of fixed amplitude and phase of the electrical component product of these two waves. This effect opens the new possibilities in re coherent optical phenomena and its application in processing of information.

The comparison of the cavity Ranman effects with scattering reabsorption of emitted photons by the ensemble of dipole-forbidden radiators is proposed in Sects. 6 and 7. Here the symmetry of the two types of resonances between the spontaneous and induced by two- and single photon transitions of three inverted radiators from the ensemble is studied. The condition for the increasing of the collective decay rate of the entangled photon pairs generated by the system relative the dipole-forbidden transition is founded. One of them corresponds to the situation when the total energy of emitted photons by two dipole-active radiators enter the two-photon resonance with the dipole-forbidden transitions of third atom. Second effect corresponds to the scattering situation, when the difference of the excited energies of two dipole-active radiators are in the resonance with the dipole-forbidden transitions of third atom. It is shown that the transfer of the excited energy between two quantum oscillators off from resonance takes place through dipole forbidden transitions of thread atom.

New coherent properties of bimodal fields can be used in lithography and quantum holography. The application possibilities of this type of coherence in holography and lithography were exposed in Sec.8. This effect usually penetrates the existing lithographic limit, and opens new possibilities for registration and reading information using entanglement (inseparability) between the photon and higher-order coherence of interacting fields.

1. Introduction

The preparation of entanglement states of photons in two or more modes was the subject of the discussions in the quantum processing systems (QPS) [1, 2, 3, 4]. These effects admit the new discussions and investigations in the fundamental description of the multi-particle entangled states named bound entanglement [5, 6]. The collective effects between the cavity photons open the conception of coherence between the bi-particles or between the Stokes and the pump and anti-Stokes photons in the cooperative Raman process (CRP). Many Refs. [1, 2, 3, 4] emphasized, that the entanglement of multi-mode particles cannot be distilled into pure entanglement under local operations of communication, a class of the multi-particle entanglement is considered useless bound entanglement for any quantum information processing. According to the authors of Ref. [7] report the first experimental preparation of a bound-entangled state. A standard scheme in quantum information theory consists of two distant parts. The Ref. [8] produced the bound entanglement in the polarization of four optical photons, produced from parametric down-conversion. Within a range of the parameter, they have shown, that this state is entangled and indistinguishable.

The realization of cooperative interaction between different subsystems in two-quantum processes is proposed in many investigations of the last time [9]. According to these concepts the multi-particle entangled states become more sophisticated than free entangled states named in many situations as an undistinguished entanglement between the particles [5, 6]. In some situations the so called bound entangled needs the new redefinition in order to describe the inseparable process which takes place between the photons belonging to same cavity modes. In such situation is better to study the quantum proprieties of collective modes, created in the nonlinear interactions between the traditional cavity modes of the electromagnetic field (EMF).

The generation of entangled photons in two- and multi-quantum processes opens the new possibilities in the realization of coherence between such a quasi-particle [10, 11, 12] . These new coherence properties of bimodal fields may be applied in quantum lithography and quantum holography. For example, in the Ref. [13] it is proposed to use the quantum entanglement for extracting holographic information about a remote $3-D$ object in a confined space which light enters, but from which it cannot escape. Light scattered from the object is detected in this confined space entirely without the benefit of spatial resolution. Quantum holography offers this possibility by virtue of the fourth-order

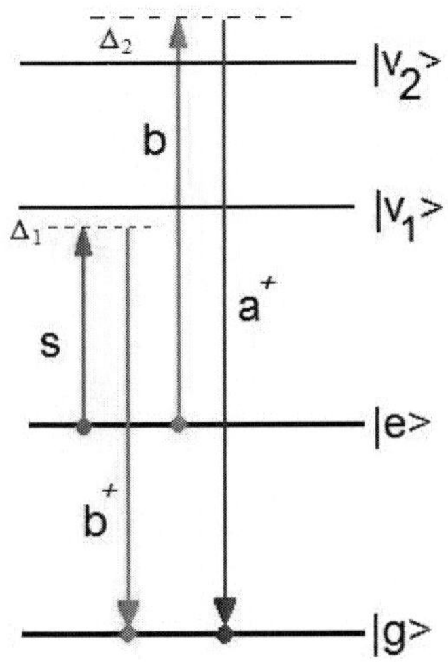

Figure 1. The scattering process with absorption and emission of anti-Stokes, pump and Stokes photons from the three cavity modes.

quantum coherence inherent in entangled beams. This new conception is based on the application of second ordered coherence function, proposed firstly by Glauber [14] and Sudarshan [15], and intensively developed in the last years. The possibility by virtue of the fourth-order quantum coherence of entangled beams was studied by Ref. [16], which proposed a two-photon analogue of classical holography. These effects establish the coherence between the collective modes in Raman emission ant two-photon lasing and open the new perspectives in quantum processing of information.

Following the new ideas about the coherence between the identical blocs of particles and observing the problems, connected with bound entangled Refs.[5, 6, 7, 8, 9], the sections 2-5 proposed the cooperative-entangled interaction between the atomic polarization and three cavity modes which correspond to the pump, Stokes and anti-Stokes photons in Raman process [11, 12]. In many cases this problem can be simplified if we take into consideration the intrinsic symmetry of bi-particle operators which describes the nonlinear interaction between the radiators and cavity filled (see Fig.1). In this situation, it is not necessary to separate individual mode, because the cooperative behavior of the particle permits us to know all about their superposition, stunning the

Hilbert vectors of such quasi-particle for example in the cavity. One of problem corresponds to the separability and reversibility of such system. In an ideal noise-free situation, any process is reversed in the sense that knowing the state of a system at a given time, it is always possible to reverse the dynamics of the system in its initial state. As in a noisy scenario, given the final state of the system, it is problematic to recover the initial state and separability between the radiators and cavity field. A specific restoration of initial states and the trapping conditions between three cavity modes and atomic subsystem are described in the Sec. 2. According to the Fig. 1 in the Sec.3 the cooperative lasing regime of three modes of cavity named, pump, Stokes and anti-Stokes, is found, and their entangled state is discussed.

Instead of the discrete cavity modes represented in the Fig. 1, in the free space we can use the excited atomic subsystem with dipole-active transitions, in order to manipulate the cooperative scattering process of dipole-forbidden transitions (see Fig. 2). The new cooperative effect between the radiators consisted of dipole active A, S and R subsystems and dipole forbidden transitions of D subsystem is established in the process of three particle spontaneous emission (see Fig. 2) A and B). This chapter is devoted to the interesting type of three particles collective spontaneous emission in which the decay rate of three atomic subsystems is proportional to the product of the numbers of atoms in each subsystem $N_sN_rN_d$ (see Fig. 2A). The quantum kinetics take into consideration both the correlations between three subsystems S, R and D and single and two-photons cooperative effects between the atoms of each sub-ensemble. The three particle cooperative interaction through the vacuum of electromagnetic field (EMF) takes place in the process of mutual influences of the single photon polarization of S and R atomic subsystem with the polarization of the D atomic ensemble relatively the two-photon transition. To understand this effect, we proposed to study the new correlation function between the polarization of three different radiators from S, R and D subsystems. In order to avoid the confusion with Dicke super-fluorescence [17], or two-photon super-fluorescence [18], in this chapter it is proposed to study the cooperative effect only between one dipole forbidden atom (D atom) and two dipole active atoms S and R (the radiators S and R don't enter in the resonance one with another). In this case only one possibility appear between such three particle ensemble : to enter into the resonance one with another according to the conservation laws: $2\omega_0 = \omega_r + \omega_s$ and $2\omega_0 = \omega_a - \omega_s$ which correspond to the two-photon resonances and scattering resonances between the three particles, respectively. The resonance be-

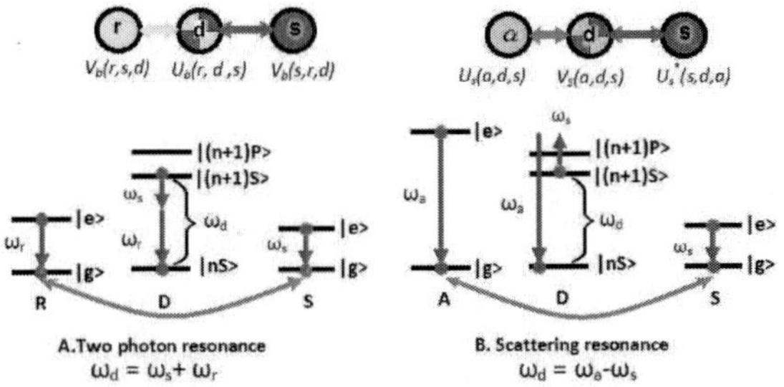

Figure 2. The resonance between the two-photon transitions of D atomic subsystem and the three dipole active systems, A, R and S is represented in figures A and B. Three atoms D, R and S are situated at relative distances r_{ds}, r_{dr} and r_{rs}. The exchange energies between the subsystems in two-photon resonance $\omega_0 = \omega_r + \omega_s$, and the scattering resonance $2\omega_0 = \omega_a - \omega_s$ are given by the expressions (35) and (37).

tween the two-photon transitions of D atomic subsystem and the two dipole active systems, A, R and S is represented in Figures 2 A and$B$. Three atoms D, R and S are situated at relative distances r_{ds}, r_{dr} and r_{rs}. The exchange energies between the subsystems in two-photon resonance $2\omega_0 = \omega_r + \omega_s$, and the scattering resonance $2\omega_0 = \omega_a - \omega_s$ are given by energetic schemes A and B. Two types of resonance between the spontaneous emissions by two- and single photon transitions of three inverted radiators from the ensemble are proposed to accelerate the collective decay rate of the entangled photons generated by the system relative the dipole-forbidden transition. One of them corresponds to the situation when the total energy of emitted photons by two dipole-active radiators enter the two-photon resonance with the dipole-forbidden transitions of third atom. Second effect corresponds to the scattering situation, when the difference between the excited energies of two dipole-active radiators are in the resonance with the dipole-forbidden transitions of third atom. These effects are accompanied with the interferences between single- and two-quantum collective transitions of three inverted radiators from the ensemble. The three particle collective decay rate is defined in the description of the atomic correlation functions.

Section 8 is devoted to applications of coherent emission of two subgroups of photons, the summer (or difference) energies of which can be regarded as a constant, so that coherence appear between the blocs of the product of two electromagnetic field strength. As it was shown recently, spatial-multi-mode broadband squeezed light can generate coherent states of the bimodal electromagnetic field, observed in photon counting experiment, the statistics of which are regular not only in time, but in the plane of observation too. The application of second order space-time coherence of light formed from entangled photons in holography and lithography follows from the definition of the quantum states of such bimodal fields. According to the proprieties of the bimodal field the coherent states are obtained in analogy with single photon coherent states. The definition of phase and amplitude of this bimodal field is given. Taking into consideration the coherent states of Stokes and anti-Stokes modes the lithographic limit of this bimodal field is found. The superposition of two vectors of bimodal field obtained in two-photon or Raman laser effect is proposed for the investigation of the holographic image of the objects. For the observation of this hologram the two-photon detection scheme is presented.

2. Symmetry between Scattering Transitions and Its Cooperative Description

In this section we represent the interaction of the stream of two-level system in Raman interaction with pump, Stokes and anti-Stokes modes of the cavity. Taking into consideration the position of virtual states $|v_1\rangle$ and $|v_2\rangle$, we neglect other possible generation modes of the electromagnetic field in the scattering process due to the large detuning from resonance. The Hamiltonian of the such two-level system connects three modes of the cavity in the Raman scattering process can be represented through the free, $\hat{H}_0$, and interaction, $\hat{H}_I$, parts by the expressions

$$\begin{aligned} \hat{H} &= \hat{H}_0 + \hat{H}_I, \\ H_0 &= \hbar\omega_a \hat{a}^\dagger \hat{a} + \hbar\omega_b \hat{b}^\dagger \hat{b} + \hbar\omega_0 \hat{R}_z + \hbar\omega_s \hat{s}^\dagger \hat{s}; \\ \hat{H}_I &= \hbar g\{\hat{b}\hat{s}^\dagger \hat{D}^\dagger + h.c.\} + \hbar\lambda\{\hat{D}^\dagger \hat{a}\hat{b}^\dagger + h.c.\}. \end{aligned} \quad (1)$$

Here the annihilation (creation) operators, $\hat{b}(\hat{b}^\dagger)$, $\hat{s}(\hat{s}^\dagger)$ and $\hat{a}\hat{a}^\dagger)$, correspond to the pump, Stokes and anti-Stokes modes, respectively, which satisfy the Bose commutation rules: $[\hat{a}_i, \hat{a}_j^\dagger] = \delta_{j,i}$, and $[\hat{a}_i, \hat{a}_j] = 0$, $j \equiv a, b, s$). The lowering, $\hat{D}^-$, and excitation, $\hat{D}^\dagger$, atomic operators are connected with atomic inversion through the commutation relation, $[\hat{D}^\dagger, \hat{D}^-] = 2\hat{D}_z$, where $\hat{D}^\dagger = \sum_{j=1}^{N_r} \hat{D}_j^\dagger$, and $\hat{D}_z = \sum_{j=1}^{N_r} \hat{D}_{zj}$ are the excitation and inversion operators of atomic subsystem.

The quantities ω_0, ω_s, ω_b and ω_a correspond to the transition frequency between the atomic levels and the frequencies of pump, Stokes and anti-Stokes modes respectively. The parameters λ and g are the Stokes and anti-Stokes coupling constants, respectively. As in the Refs.[11] we consider that the coupling constant λ (g) denotes the coupling between the Stokes (anti-Stokes) mode. The atomic collective polarization operator depends on the number of the radiators, N_r $\hat{D}^{\dagger} = \hat{c}^{\dagger}\sqrt{N_r - \hat{c}^{\dagger}\hat{c}}$ and $\hat{D}^{-} = \sqrt{N_r - \hat{c}^{\dagger}\hat{c}\hat{c}}$), and for a large number of atoms in the ground state, $N_r \gg \langle \hat{c}^{\dagger}\hat{c}\rangle$, these operators can be regarded as a boson creation and annihilation operators $\hat{R}^{\dagger} \simeq \hat{c}^{\dagger}\sqrt{N_r}$, $\hat{R}^{-} \simeq \sqrt{N_r}\hat{c}$. For molecular vibration, described by above boson operators, when the pump is very strong, so that its field operator can be described by c-number operator, the Schrdinger equation can be solved exactly [19]. In order to solve the Schrdinger equation under weak pump approximation the authors of the Ref.[19] used the perturbation approach. In same short-time approximation, taking into consideration the intrinsic symmetry between the bi-Boston operator, we proposed a new exact solution. Regarding the atomic subsystem as a pump beam of the Stokes and anti-Stokes modes into the resonator, we describe this lasing generation using the collective photon modes.

Let us study short-time approximation, when the atomic or photon losses from the optical cavity is neglected. Following the method proposed in the Ref.[20], we may define the $N+1$ vector operators of the atomic ensemble in the Raman interaction with three cavity modes, described by the Hamiltonian (1)

$$|\psi(t)\rangle = \sum_{m=N/2}^{N/2} \alpha_m \hat{X}_m |\varphi_{ph}\rangle, \tag{2}$$

where $\hat{X}_m(t) = \exp[-i\hat{H}_I t]\ |j,m\rangle$. Taking into consideration that the action of the operators $\hat{D}^{\dagger}$ and $\hat{D}^{-}$ on the collective state $\hat{D}^{\dagger}|j,m\rangle = \sqrt{(j+m+1)(j-m)}\ |j,m+1\rangle$, $\hat{D}^{-}|j,m\rangle = \sqrt{(j+m)(j-m+1)}|j,m-1\rangle$, we obtain the system of N_r+1 equations for these operator-vectors

$$\begin{aligned} \frac{d}{dt}\hat{X}_m(t) &= -i\sqrt{(j+m+1)(j-m)}\hat{X}_{m+1}(t)[g\hat{b}\hat{s}^{\dagger} + \lambda\hat{a}\hat{b}^{\dagger}] \\ &- i\sqrt{(j+m)(j-m+1)}\hat{X}_{m-1}(t)[g\hat{s}\hat{b}^{\dagger} + \lambda\hat{b}\hat{a}^{\dagger}], \\ &\quad m = -j, -j+1, ..., j+1, j. \end{aligned} \tag{3}$$

If we have the single atom in the cavity the solution of two equations of the system of two equations (3) is described by the set of two operator-vectors

$$\begin{aligned}\hat{X}_{0.5}(t) &= |e\rangle \cos(\hat{\Omega}t) - i\,|g\rangle\, [g\hat{s}\hat{b}^\dagger + \lambda\hat{b}\hat{a}^\dagger]\hat{\Omega}^{-1}\sin(\hat{\Omega}t);\\ \hat{X}_{-0.5}(t) &= |g\rangle \cos\left[\sqrt{\hat{\Omega}^2 - g^2(\hat{s}^\dagger\hat{s} - \hat{b}\hat{b}^\dagger) - \lambda^2(\hat{b}^\dagger\hat{b} - \hat{a}\hat{a}^\dagger)}t\right]\\ &- i\,|e\rangle\, [g\hat{b}\hat{s}^\dagger + \lambda\hat{a}\hat{b}^\dagger]\left[\sqrt{\hat{\Omega}^2 - g^2(\hat{s}^\dagger\hat{s} - \hat{b}\hat{b}^\dagger) - \lambda^2(\hat{b}^\dagger\hat{b} - \hat{a}\hat{a}^\dagger)}\right]^{-1}\\ &\times \sin\left[\sqrt{\hat{\Omega}^2 - g^2(\hat{s}^\dagger\hat{s} - \hat{b}\hat{b}^\dagger) - \lambda^2(\hat{b}^\dagger\hat{b} - \hat{a}\hat{a}^\dagger)}t\right], \qquad (4)\end{aligned}$$

where $\hat{\Omega} = \sqrt{[g\hat{b}\hat{s}^\dagger + \lambda\hat{a}\hat{b}^\dagger][g\hat{s}\hat{b}^\dagger + \lambda\hat{b}\hat{a}^\dagger]}$. The ground $|g\rangle$ and excited $|e\rangle$ states of the atom are described by two $|1/2, 1/2\rangle$ and $|1/2, -1/2\rangle$ angular momentum vectors for $j = 1/2$. It is not so difficult to observe, that the substitution of the vectors in the solution (2) is reduced to Jaynes-Cummings model with non-linear interaction of an atom with the field. We are interested in the simplification and symmetries of the bi-boson interaction of atoms with scattering field described by the Hamiltonian (1). From commutation of the bi-boson field operators,$[g\hat{s}\hat{b}^\dagger + \lambda\hat{b}\hat{a}^\dagger, g\hat{b}\hat{s}^\dagger + \lambda\hat{a}\hat{b}^\dagger]$=$g^2(\hat{b}^\dagger\hat{b} - \hat{s}\hat{s}^\dagger) + \lambda^2(\hat{a}^\dagger\hat{a} - \hat{b}\hat{b}^\dagger)$, it is not difficult to observe, that, when the constant interaction of atoms with Stokes and anti-Stokes modes coincide, $g = \lambda$, the new operators, belonging to angular momentum $SU(2)$ symmetry, could easily be introduced: $\hat{L}_z = \hat{a}^\dagger\hat{a} - \hat{s}^\dagger\hat{s}, \quad \hat{L}^- = \sqrt{2}(\hat{b}\hat{s}^\dagger + \hat{a}\hat{b}^\dagger), \quad \hat{L}^+ = \sqrt{2}(\hat{s}\hat{b}^\dagger + \hat{b}\hat{a}^\dagger)$. The commutation relations between these operators are similar to the commutators of the collective atomic operators: $[\hat{L}^+, \hat{L}^-] = 2\hat{J}_z$ and $[\hat{L}_z, \hat{L}^\pm] = \pm\hat{L}^\pm$. According to this description the Hamiltonian (1),

$$\hat{H} = \hbar(\omega_a - \omega_b)\hat{L}_z + \hbar\omega_0\hat{R}_z + \frac{\hbar g}{\sqrt{2}}\{\hat{L}^-\hat{R}^\dagger + \hat{R}^-\hat{L}^+\}, \qquad (5)$$

describes the interaction between two cooperative subsystems atoms and photons. The operator of inversion doesn't contain the pump boson operators, so that the pump mode operators is not represented in the inversion. In this Hamiltonian two Bloch Vectors are conserved. First of them is the atomic Bloch vector $j(j+1) = \hat{D}_z^2 + \{\hat{D}^-\hat{D}^\dagger + \hat{D}^-\hat{D}^+\}$. The second corresponds to the three cavity modes represented by the vector $l(l+1) = \hat{L}_z^2 + \{\hat{L}^-\hat{L}^\dagger + \hat{L}^-\hat{L}^+\}/2$. In this case the integer number l represents the total photon in the system. The conversion operator of the photon between the Stokes and anti-Stokes Modes takes place through the pump mode.

Let us study the connection between the angular momentum states for the photon subsystem $|l,m\rangle$, $m=-l,...,+l$, and the three Fock states of the cavity modes. Taking into consideration the action of the operator $\hat{L}_z$ on this state is $\hat{L}_z|l,m\rangle = m|l,m\rangle$, $\hat{L}^+|l,m\rangle = \sqrt{(l-m)(l+m+1)}|l,m\rangle$ and $\hat{L}^-|l,m\rangle = \sqrt{(l-m+1)(l+m)}\;|l,m\rangle$ we can found these actions on the three boson states of the cavity field. In the literature it is well known relation between two-boson correlated states and angular momentum vectors $|l,m\rangle = |n_t-p\rangle_b|p\rangle_s$ where$|k\rangle_b$ and $|i\rangle_s$ are the Fock states for two boson modes in the scattering process in which only pump and Stokes modes $\hat{L}^+ = \hat{b}\hat{s}^\dagger$, $\hat{L}^- = \hat{b}^\dagger\hat{s}$, $k+1 = n_f$ is take into consideration. In the situation when we take into consideration three modes of the cavity $\hat{L}^- = \sqrt{2}(\hat{b}\hat{s}^\dagger + \hat{a}\hat{b}^\dagger)$ and $\hat{L}^+ = \sqrt{2}(\hat{s}\hat{b}^\dagger + \hat{b}\hat{a}^\dagger)$, it is necessary to found the correspondence between three boson states $|l,m\rangle = |i\rangle_s|k\rangle_b|p\rangle_a$ with the fixed total number of photons $i+p+k=n_f$ and $|l,m\rangle$ states. Let us start with simple model, in which the single photon is in anti-Stokes mode $|1,1\rangle$. The action of operator $\hat{L}^- = \sqrt{2}(\hat{b}\hat{s}^\dagger + \hat{a}\hat{b}^\dagger)$. This action corresponds to the action of this operator to the three boson state $|0\rangle_s|0\rangle_b|1\rangle_a$ in which photon pass into pump mode $\hat{L}^-|0\rangle_s|0\rangle_b|1\rangle_a = \sqrt{2}$ $0\rangle_s|1\rangle_b|0\rangle_a$. The second action of this operator transform it in Stokes photon: $\hat{L}^-|0\rangle_s|1\rangle_b|0\rangle_a = \sqrt{2}|1\rangle_s|0\rangle_b|0\rangle_a$. Here we observe the good correspondence between the angular momentum and three boson states $|1,1\rangle$, $|1,0\rangle$, $|1,-1\rangle$ and $|0\rangle_s|0\rangle_b|1\rangle_a$, $|0\rangle_s|1\rangle_b|0\rangle_a$, $|1\rangle_s|0\rangle_b|0\rangle_a$. The correspondence between these two basis of vectors becomes more tractive if we consider the large number of photons in anti-Stokes mode. Indeed if we represent the angular momentum state through the state $|l,l-p\rangle$, we can found the following superposition. For $p=1:$ —$0\rangle_s|0\rangle_b|l\rangle_a$; $|0\rangle_s|1\rangle_a|l-1\rangle_a$; For $p=2$ and $p=3$ the superposition between two Fock states: $|l,l-2\rangle = \{|1\rangle_s|0\rangle_b|l-1\rangle_a + \sqrt{2(l-1)}|0\rangle_s|2\rangle_b|l-2\rangle_a\}/\sqrt{2l-1}$ and $|l,l-3\rangle = [\sqrt{3}|1\rangle_s|1\rangle_b|l-2\rangle_a + \sqrt{2(l-2)}|0\rangle_s|3\rangle_b|l-3\rangle_a]/\sqrt{2l-1}$; For $p=4$ and $p=5$ the superposition of three Fock states of the photons:$|l,l-4\rangle = \{\sqrt{3}|2\rangle_s|0\rangle_b|l-2\rangle_a + 2\sqrt{3(l-2)}|1\rangle_s|2\rangle_b|l-3\rangle_a + 2\sqrt{(l-2)(l-3)}|0\rangle_s|4\rangle_b|j-4\rangle_a\}/\sqrt{(2l-1)(2l-3)}$ and $|l,l-5\rangle = \{\sqrt{5\times 3}|2\rangle_s|1\rangle_b|l-3\rangle_a$ +// $2\sqrt{5(l-3)}|1\rangle_s|3\rangle_b|l-4\rangle_a + 2\sqrt{(l-3)(l-4)}|0\rangle_s|5\rangle_b|l-5\rangle_a\}/\sqrt{(2l-1)}\sqrt{(2l-3)}$; for $p=6$ and $p=7$ we have the combination of four terms of three particle states:

$$\begin{aligned}|l,\quad l-6\rangle &= \{\sqrt{5\times 3}|3\rangle_s|0\rangle_b|l-3\rangle_a + \sqrt{6\times 5\times 3(l+3)}|2\rangle_s|2\rangle_b|l-4\rangle_a\\ &+ \sqrt{6\times 5\times 2(l-3)(l-4)}|1\rangle_s|4\rangle_b|l-5\rangle_a\\ &+ 2\sqrt{2(l-3)(l-4)(l-5)}|0\rangle_s|6\rangle_b|l-6\rangle_a\}/\sqrt{(2l-1)(2l-3)(2l-5)},\end{aligned}$$

and

$$\begin{aligned}|l,\quad l-7\rangle &= \{\sqrt{7\times 5\times 3}|3\rangle_s|1\rangle_p|l-4\rangle_a + \sqrt{7\times 5\times 3\times 2(l-4)}\\ \times&\quad |2\rangle_s|3\rangle_p|l-5\rangle_a + 2\sqrt{7\times 3(l-4)(l-5)}|1\rangle_s|5\rangle_p|l-6\rangle_a\\ +&\quad 2\sqrt{2(l-4)(l-5)(l-6)}|0\rangle_s|7\rangle_p|l-7\rangle_a\}/\sqrt{(2l-1)(2l-3)(2l-5)}.\end{aligned}$$

This procedure can be continued till ground vector $|l, -l\rangle$. This procedure of the construction of the eigan states of the conserved angular momentum of the three mode cavity field can be realized acting consecutively to the "ground" Stokes state $|l\rangle_s|0\rangle_p|0\rangle_a$ with the operator $\{\hat{L}^+\}^k$.

In order to observe this let us take the an atom in interaction with a single photon prepared in anti-Stokes state, and zero photons in the pump and Stokes modes. In this case we have three cooperative vectors in the Hilbert space of bi-Boston subsystem: $|1,1\rangle$, $|1,0\rangle$, —$1,-1\rangle$ one photon each anti-Stokes, pump and Stokes modes respectively. The above two solutions look simpler in the new representation

$$\begin{aligned}|\psi(t)\rangle &= \{\alpha[|e\rangle\cos(gt\sqrt{\hat{L}^-\hat{L}^+/2}) - i\,|g\rangle\,\hat{D}^+\frac{1}{\sqrt{\hat{D}^-\hat{D}^+}}\sin(gt\sqrt{\hat{L}^-\hat{L}^+/2})]\\ &+ \beta[|g\rangle\cos(gt\sqrt{(\hat{L}^+\hat{L}^-/2})\\ &- i\,|e\rangle\,\hat{L}^-\left[\sqrt{\hat{L}^+\hat{L}^-}\right]^{-1}\sin(gt\sqrt{(\hat{L}^+\hat{L}^-/2)t})]\}\\ &\times \{C_1|1,1\rangle + C_0|1,0\rangle + C_{-1}|1,-1\rangle\}. \qquad (6)\end{aligned}$$

Here we have considered that in the initial time moment $t = 0$ the atom and bi-boson field was prepared in the superposition $\alpha\,|e\rangle + \beta\,|g\rangle$ and $C_1|1,1\rangle + C_0|1,0\rangle + C_{-1}|1,-1\rangle$. Taking into consideration the action of the cooperators on the bi-photon states we can find the conditional time of the dynamical trapping for which in the special preparation of factored initial state the wave function of atom and field remains factorized. For example the superposition factorized state $\psi(t) = [\alpha\,|e\rangle + \beta\,|g\rangle][C_1|1,1\rangle + C_{-1}|1,-1\rangle]$ for the discrete flying time, $t_n = n\pi/g$, $(n = 1,\ 2,\ ...)$ into same state $|\psi(t)\rangle = [\alpha\,|e\rangle + (-1)^n\beta\,|g\rangle][C_1|1,1\rangle + (-1)^n C_{-1}|1,-1\rangle]$ changing only the phase of the factorized wave function.

The new Hamiltonian (5) permits the exact solution for two undistinguished atoms. According to the definition of angular momentum vectors the set of three states $\{|1,m\rangle, m = -1,0,1\}$ correspondents to the collective three level states $|e\rangle = |e_1\rangle\,|e_2\rangle$; $|g\rangle = |g_1\rangle\,|g_1\rangle$; $|i\rangle = [|e_1\rangle\,|g_2\rangle + |e_2\rangle\,|g_1\rangle]/\sqrt{2}$, which can be excited and lowered by alder operators: $\hat{D}^\dagger = \sqrt{2}(|e\rangle\,\langle i| + |i\rangle\,\langle g|)$; $\hat{D}^- = \sqrt{2}(|i\rangle\,\langle e| + |g\rangle\,\langle i|)$. As in the bi-photon cooperative state the inversion operator doesn't contain intermediate state $\hat{D}_z = |e\rangle\langle e| - |g\rangle\langle g|$. The solution of Schrdinger equation for three level atom can be represented in the superpositional states.

3. Open System Cooperative Effects between the Photon Modes Stimulated by the Stream of the Two Level Atom in the Scattering Resonance

Let us analyse the situation, when the flying time is shorter than the vacuum Rabi frequency, obtained in the last section. In this case we can rewrite the Heisenberg equation for atomic operators, introducing the losses of atoms from the cavity. Indeed, taking into consideration the atomic source of excited atoms in the cavity and the Hamiltonian (5), we propose th following Heisenberg equations for the atomic operators similar to laser description in Ref.[24]

$$
\begin{aligned}
\frac{d\hat{D}_j^{\dagger}(t)}{dt} &= (i\omega_0 - \gamma)\hat{D}_j^{\dagger}(t) \\
&- i\sqrt{2}g\hat{D}_{zj}(t)\hat{L}^{\dagger}(t)\exp[ik_0x_j], \\
\frac{d\hat{D}_j^{-}(t)}{dt} &= (-i\omega_0 - \gamma)\hat{D}^{-}(t) + i\sqrt{2}g\hat{D}_{zj}(t)\hat{L}^{-}(t)\exp[ik_0x_j], \\
\frac{d\hat{D}_{zj}(t)}{dt} &= -\gamma_p[\hat{D}_z(t) - \frac{1}{2}] + i\frac{g}{\sqrt{2}} \\
&\times \left\{\hat{D}_j^{-}(t)\hat{L}^{\dagger}(t)\exp[ik_0x_j] - \hat{L}^{-}(t)\hat{D}_l^{+}(t)\exp[-ik_0x_j]\right\}.
\end{aligned}
$$

Here $\gamma^{-1} = \Lambda/v$ is flying time expressed through the atomic velocity, v, and the cavity length Λ , γ_p is the atomic pump rate in the cavity. Neglecting the Doppler effect in the atomic distribution, we may approximate the cavity pump and losses with some parameter, $\gamma \simeq \gamma_p$. The first two equations describe the attenuation of the positive (negative) atomic polarization part due to shot time living time of the radiator in the cavity. The third equation describes the inversion modification of the atom during the pump and emission process. Considering that the lifetime of atoms in the cavity is shorter than the nutation period of the atom in the cavity, we may adiabatically eliminate the atomic variable from the chain of the equation for collective operators of the cavity EMF. In order to obtain the master equation of cavity *bi*-modes, described by the new collective operators, let us introduce the generalized operator $O(t)$ of photon subsystem $\hat{O}(t)$

$$\frac{d}{dt}\left\langle\hat{O}(t)\right\rangle = i\tilde{\omega}\left\langle\left[\hat{L}_z(t),\hat{O}(t)\right]\right\rangle$$
$$+i \quad \frac{g}{\sqrt{2}}\sum_{l=1}^{N} g\{\langle\hat{D}_l^-(t)[\hat{O}(t),\hat{L}^+(t)]\rangle\exp[ik_0x_l] \tag{7}$$
$$- \quad \langle[\hat{L}^-(t),\hat{O}(t)]\hat{D}_l^{;\dagger}(t)\rangle\exp[-ik_0x_l]\}.$$

If the atomic stream is prepared in the coherent states [22], we can eliminate the polarization operators from the generalized equation (7), using the method of Ref.[23]. We are interested in the developing the method, which takes into consideration the quantum fluctuation in the higher order of perturbation theory. For this we consider only the simplest situation, in which the atomic stream is prepared in the excited (or ground) state. Let us represent the positive and negative part of the atomic polarization in through the source and free parts, $\hat{D}_j^{\pm}(t) = \hat{D}_j^{\pm f}(t) + \hat{D}_j^{\pm s}(t)$, so that the free $\hat{D}_j^{\pm f}(t)$ and sources $\hat{D}_j^{\pm s}(t)$ parts are described by the expressions

$$\hat{D}_l^{\pm f}(t) = \hat{D}_l^{\ \pm f}(0)\exp[\pm i\omega_0 t - \gamma t],$$
$$\hat{D}_l^{\pm s}(t) = \mp i \quad \sqrt{2}g\int_0^t d\tau\exp[\pm(i\omega_0\tau + ik_0x_l) - \gamma\tau]\hat{D}_{zl}(t-\tau)\hat{L}^{\pm}(t-\tau).\tag{8}$$

We can eliminate the free part of excitation operator and de-excitation operators representing the correlations of the right hand site of the generalized equation (7) in anti-normal form, so that the actions are $\hat{D}_f^{\ \dagger}(0)|e_1,..e_N\rangle = 0$, $\langle e_1,..e_N|\hat{D}_f^{\ -}(0) = 0$. After the substitution of the Born-Marcoff approximated expression (8)

$$\hat{D}_l^{\pm s}(t) = \pm i\sqrt{2}g\exp[\pm ik_0x_l]\hat{D}_{zl}(t)\hat{L}^{\pm}(t)/[\pm i(\omega_0-\tilde{\omega})-\gamma] \tag{9}$$

in equation antinormal form of equation (7) we obtain

$$\frac{d\langle\hat{O}(t)\rangle}{dt} = i\tilde{\omega}\langle[\hat{L}_z(t),\hat{O}(t)]\rangle$$
$$+ \sum_{l=1}^{n}|g|^2\{\frac{\langle\hat{D}_{zl}(t)\hat{L}^-(t)[\hat{L}^+(t),\hat{O}(t)]\rangle}{i(\tilde{\omega}-\omega_0)-\gamma}$$
$$+ \frac{\langle[\hat{L}^-(t),\hat{O}(t)]\hat{L}^+(t)\hat{D}_{zl}(t)\rangle}{i(\tilde{\omega}-\omega_0)+\gamma}. \tag{10}$$

Considering the conversion rate between pump, Stokes and anti-Stokes photons as a small parameter $g^2/\gamma < 1$, we can use the Born-Marcoff approximation in all orders of perturbation theory. Here $1/\tau = Ng^2$ is the conversion rate. In such approximation we can realize the stabilization of the of the conversion rate of Stokes photons in anti-Stokes. In this Raman process the new operators describe the cooperative transformation of the Stokes, pump and anti-Stokes photons between them. These processes take place with conservation of the total vector: $L^2 = l(l+1)$. For example, considering that in the initial time moment $t = 0$ all photons are prepared in Stokes mode , $l = n$, we find induced conversion of this number of photons in anti-Stokes photons. In the resonance case, when $\tilde{\omega} - \omega_0$ and $\gamma = \gamma_p$, let us continue to eliminate the atomic operators from the equation (10), representing the atomic inversion through the solution of the third equation of the system (8)

$$\begin{aligned} \hat{D}_{zj}(t) &= \frac{1}{2} + \frac{ig}{\sqrt{2}\gamma}\{[\hat{D}_j^{-s}(t) + \hat{D}_j^{-f}(t)]\hat{L}^+(t)\exp[ik_0x_j] \qquad (11) \\ &- \hat{L}^-(t)[\hat{D}_j^{+s}(t) + \hat{D}_j^{-f}(t)]\exp[-ik_0x_j]\}; \\ \hat{D}_l^{\pm s}(t) &= \pm i\sqrt{2}g\exp[\pm ik_0x_l]\hat{D}_{zl}(t)\hat{L}^\pm(t)/[\pm i(\omega_0 - \tilde{\omega}) - \gamma]. \end{aligned}$$

In the first order of perturbation theory, we can approximate the inversion with its initial value, $\hat{D}_{zj}(t) = 0.5$, in the right hand site of the generalized equation (10). In this case we observe, that the conversion of photons from Stokes to anti-Stokes modes (through the pump cavity mode) takes place like in the Dicke superradiance. If we consider instead of $\hat{O}(t)$ the conversion operator $\hat{L}_z(t)$ in the generalized equation (10), according to the conservation law, $\hat{L}^-\hat{L}^\dagger = l(l+1) - \hat{L}_z^2 - \hat{L}_z$, we obtain the following expression

$$\frac{d\langle\hat{L}_z(t)\rangle}{dt} = \frac{1}{\tau}\left\{l(l+1) - \langle\hat{L}_z^2(t)\rangle - \left\langle\hat{L}_z(t)\right\rangle\right\}.$$

For the large values of number of Stokes photon n we can neglect the quantum fluctuation of conversion operator e $\langle\hat{L}_z^2(t)\rangle \simeq \langle\hat{L}_z(t)\rangle^2$. Taking into consideration the initial state $\left\langle\hat{L}_z(0)\right\rangle = -n$, it is obtained the superradiant solution for conversion :$\langle\hat{L}_z(t)\rangle = -n\tanh[(t - t_0)/\tau_r]$, where$\tau_r = \tau/n$ and $t_0 = \tau_r\ln[2n]$ are the cooperative and delay time in the theory of superradiance [17]. In the case, when all photons are prepared in Stokes mode, we have the conversion rate from Stokes to pump and from Pump to anti-stokes photons, so that, when the Bloch vector pass through the horizontal plane of Bloch sphere the correlation between the Stokes, pump and anti-Stokes photons becomes maximally entangled $\langle \hat{L}^{\dagger}\hat{L}^{-}\rangle \simeq l^2$, where $\hat{L}^{\dagger}\hat{L}^{-} = 2(\hat{s}\hat{b}^{\dagger} + \hat{b}\hat{a}^{\dagger})(\hat{b}\hat{s}^{\dagger} + \hat{a}\hat{b}^{\dagger})$.

This process of photon conversion takes place till all photons are transformed totally into anti-Stokes photons. It is not difficult to observe, that, when the atoms are prepared in the ground state and the photons are prepared in anti-Stokes mode, the photon conversion takes place inversely: All photons are transformed from anti-Stokes to Stokes modes during the cooperative excitation of the system. In this case in equations (10) we must do the replacement $\hat{L}^{+}(t) \leftrightarrow \hat{L}^{-}(t)$ and the atomic envision can be approximated with initial value $-1/2$. According to this, the mean value of the photon conversion operator becomes $\langle \hat{L}_z(t)\rangle = n\tanh[(t-t_0)/\tau_r]$. These finite processes are possible because in comparison with the spectrum of the Fock states for single Boson mode the new cooperative operators have the finite Hilbert vectors and states $-n \leq m \leq n$.

The consecutive decomposition on the small parameter, $\epsilon = g^2/\gamma^2$, can be applied, using the results of Refs. [11, 12]. This method proposes to substitute the first order of inversion $\hat{D}_{zl} \simeq 1/2$ in the solution for polarization (8). After the substitution of this approximation, $\hat{D}^{\pm}_{l,s}(t) = \pm i(\sqrt{2})^{-1}g\exp[\pm ik_0x_l]\hat{L}^{\pm}(t)/[\pm i(\omega_0-\tilde{\omega})-\gamma]$, instead of the expression for the polarization in the right hand side of the operator of inversion (11), we obtain

$$\begin{aligned}\hat{D}_{zj}(t) &\simeq \frac{1}{2} + \frac{ig}{\sqrt{2}\gamma}\{[i(\sqrt{2})^{-1}g\exp(-ik_0x_j)\hat{L}^{-}(t)/[i(\omega_0-\tilde{\omega})+\gamma] \\ &+ \hat{D}_j^{-f}(t)]\hat{L}^{+}(t)\exp[ik_0x_j] - H.c\}. \qquad (12)\end{aligned}$$

In order to improve the decomposition procedure,we must continue to eliminate the free part of the polarization operators $\hat{D}_j^{-f}(t)$ and$\hat{D}_j^{+f}(t)$ from the master equation (10), using the anti-normal arrangement of the operators in correlation functions. In such a way we obtain the consecutive decomposition of the right hand of the generalized equation proposed in Refs.[11, 12]. In the next section we revised this method of decomposition. The new small parameter for the decomposition of the right hand side of the generalized equation is proposed in the section 4

4. Decomposition Method for Master Equation of Three Cavity Mode in Raman Process

Considering that the life time of the atoms in the cavity increases, we must take into consideration the higher order of perturbation parameter ϵ. In this case the Bloch vector is stopped on the azimuthal angle ϕ which determine the stabilization of coherent state between the photons of three cavity modes Stokes, pump and anti-Stoke. In the right hand of the equation for inversion, we can introduce the Born-Marcoff expression for polarization operators (8) in the right hand site of the operator of the atomic inversion (11) so that it can be solved relatively the operator of inversion

$$\begin{aligned} \hat{D}_{zl}(t) &= \left\{ \frac{1}{2} + \frac{ig}{\sqrt{2}\gamma} [\hat{D}_l^{-f}(t)\hat{L}^+(t)\exp[-ik_0x_l] - H.c.)] \right\} \\ &\times \frac{1}{1 + 2g^2\gamma^{-2}\hat{L}^-(t)\hat{L}^\dagger(t)}, \\ \hat{D}_{zl}(t) &= [\hat{D}_{zl}(t)]^+. \end{aligned} \tag{13}$$

Here we consider, that in the Born-Marcoff approximation the operator combination of field and atomic operators $i[\hat{D}_l^{-f}(t)\hat{L}^+(t)\exp[-ik_0x_l] - H.c.]$ commutes with the expression $1/(1 + 2g^2\gamma^{-2}\hat{L}^-(t)\hat{L}^+(t))$. In this case we used both representations 13) in the right hand side of equation (10). Introducing this solution in expression (10), we obtain the new decomposition, in which the small parameter, ϵ, appears also and in the denominator. This decomposition permits us to eliminate the free part of the operators $\hat{D}_l^{-f}(t)$ from the second and third terms of the equation (10)

$$\begin{aligned} \frac{d}{dt}\langle \hat{O}(t)\rangle &= i\tilde{\omega}\langle [\hat{L}_z(t), \hat{O}(t)]\rangle \\ &+ \frac{1}{2\tau}\left\langle \frac{1}{1 + 2g^2\gamma^{-2}\hat{L}^-(t)\hat{L}^+(t)} \hat{L}^-(t)[\hat{O}(t), \hat{L}^+(t)]\right\rangle + \frac{ige^{ik_0x_l}}{\sqrt{2}\gamma\tau} \\ &\times \left\langle \hat{L}^-(t)D_l^{+f}(t)\frac{1}{1 + 2g^2\gamma^{-2}\hat{L}^-(t)\hat{L}^+(t)}\hat{L}^-(t)[\hat{O}(t), \hat{L}^+(t)]\right\rangle \\ &+ H.c. \end{aligned} \tag{14}$$

The difficult problem appears with elimination of the free parts of atomic operators from this expression. For elimination of this part $\hat{D}_l^{\dagger f}(t)$ (or $\hat{D}_l^{-f}$) from the equation (14) we must permute the operator from the left (right hand) to the right hand (left hand). The physical aspect of this procedure is connected with the increasing quantum fluctuations with the decreasing of the inversion in the process of the transition of atoms from the excited to the ground states. In this case the absorption of Stokes photon is accompanied by the emission of the pump photons. In the similar situation the absorption of pump photon with the emission of anti-Stokes one is possible. This process is connected with the re-excitation of atoms, in which we take into consideration their re-excitations. This effect is responsible for the stabilization of the multi-photon induced processes in the new modes. This steady state corresponds to the relaxation of Bloch vector in the one of the states situated on the Bloch sphere.

In the second order of perturbation theory, we can introduce the diffusion coefficient, $1/\Theta = g^4 N/[\gamma^3] = (1/\tau)\epsilon$, which increases with the increasing the lifetime of the atoms in the cavity. We observe, that the divergence in the decomposition on the small parameter, ϵ, increases with the increasing of the number of photons, " n" of the collective modes, described by the right hand side of equation (14). In this situation the number of terms in the second order of perturbation theory becomes proportional to square number of cooperative photons, "n^2".

$$
\begin{aligned}
\frac{d}{dt}\langle \hat{O}(t)\rangle &= i\tilde{\omega}\langle[\hat{L}_z(t), \hat{O}(t)]\rangle \\
&+ \frac{1}{2\tau}\langle\frac{1}{1+2\epsilon\hat{L}^-(t)\hat{L}^\dagger(t)}\hat{L}^-(t)[\hat{O}(t), \hat{L}^\dagger(t)]\rangle \\
&- \frac{1}{\Theta}\langle\hat{R}_{zl}(t)\hat{L}^-(t)[\hat{D}^\dagger(t), \frac{1}{1+2\epsilon\hat{L}^-(t)\hat{L}^\dagger(t)}\hat{L}^-(t)[\hat{O}(t), \hat{L}^\dagger(t)]]\rangle \\
&+ H.c. \qquad (15)
\end{aligned}
$$

With increasing of the cooperative number (connected with the number of photons in the cavity), the expression of the denominator also increases. So on the right hand side of this expression the number of therms also increases. The parameter $x/(1+x)$ achieves the maximal value "1", when the mean value of the operator $x = \langle 2\epsilon\hat{L}^-(t)\hat{L}^\dagger(t)\rangle$ increases. In this situation we must take into consideration a larger number of therms in the right hand side of the expression (15). Here $\hat{L}^-(t)\hat{L}^\dagger(t)$ is considered proportional to the n^2 as in the Dicke super-radiance. In comparison with the Refs.[11, 12], where the function $x/(1+x)$ was approximated by the expression $x - x^2$, in this section we decompose the right hand side of the master equation in the serial function relative the new small parameter $\delta = x/(1+x) < 1$, which remain smaller than one, when x achieved value $x = 1$. Indeed intruding this expression again on the right hand side of the expression (15), we can obtain the new decomposition on the small parameter, $\epsilon/(1+\epsilon\hat{L}^{-}(t)\hat{L}^{\dagger}(t))$

$$\begin{aligned}
\frac{d}{dt}\langle\hat{O}(t)\rangle &= i\tilde{\omega}\langle[\hat{L}_z(t),\hat{O}(t)]\rangle \\
&+ \frac{1}{2\tau}\langle\frac{1}{1+2\epsilon\hat{L}^{-}(t)\hat{L}^{\dagger}(t)}\hat{L}^{-}(t)[\hat{O}(t),\hat{L}^{\dagger}(t)]\rangle \\
&- \frac{1}{2\tau}\epsilon\langle\hat{L}^{-}(t)[\hat{L}^{\dagger}(t),\frac{1}{1+2\epsilon\hat{L}^{-}(t)\hat{L}^{\dagger}(t)}\hat{L}^{-}(t)[\hat{O}(t),\hat{L}^{\dagger}(t)]]\rangle \\
&+ \frac{1}{2\tau}\epsilon^2\langle\hat{L}^{-}(t)[\hat{L}^{+}(t),\frac{1}{1+2\epsilon\hat{L}^{-}(t)\hat{L}^{\dagger}(t)}\hat{L}^{-}(t)[\hat{L}^{\dagger}(t), \\
&\quad \frac{1}{1+2\epsilon\hat{L}^{-}(t)\hat{L}^{\dagger}(t)}\hat{L}^{-}(t)[\hat{O}(t),\hat{L}^{\dagger}(t)]]]\rangle \\
&+ \ldots+\frac{1}{2\tau}(-\epsilon)^n\langle\hat{L}^{-}(t)[\hat{L}^{+}(t),\frac{1}{1+2\epsilon\hat{L}^{-}(t)\hat{L}^{\dagger}(t)}\hat{L}^{-}(t)\ldots[\hat{L}^{\dagger}(t), \\
&\quad \frac{1}{1+2\epsilon\hat{L}^{-}(t)\hat{L}^{\dagger}(t)}\hat{L}^{-}(t)[\hat{O}(t),\hat{L}^{\dagger}(t)]\ldots]\rangle \\
&+ H.c. \qquad (16)
\end{aligned}$$

Using the interaction picture $Tr\{\hat{O}(t)\hat{W}(0)\}=Tr\{\breve{O}(t)\hat{W}(t)\}$, we pass from the generalized operator $\hat{O}(t)$ to the density matrix using the interaction picture $Tr\{\hat{O}(t)\hat{W}(0)\}=Tr\{\breve{O}(t)\hat{W}(t)\}$. The following expression for master equation is obtained after the permutations

$$\begin{aligned}
\frac{d}{dt}\hat{W}(t) &= -\frac{1}{2\tau}[\hat{W}(t)\hat{L}^{-}\frac{1}{1+2\epsilon\hat{L}^{-}\hat{L}^{\dagger}},\hat{L}^{\dagger}] \\
&- \frac{1}{2\tau}\sum_{n=1}^{\infty}(-1)^n\epsilon^n[\ldots[[\hat{W}(t)\hat{L}^{-},\hat{L}^{+}]\frac{1}{1+2\epsilon\hat{L}^{-}\hat{L}^{\dagger}}\hat{L}^{-},\hat{L}^{+}]\ldots \\
&\times \frac{1}{1+2\epsilon\hat{L}^{-}\hat{L}^{\dagger}}\hat{L}^{-},\hat{L}^{+}]+H.c. \qquad (17)
\end{aligned}$$

Both expressions (16) and (17) are represented by infinite series on the small parameter δ. We may formally represent Eq. (17) in the compact form

$$\begin{aligned}
\frac{d}{dt}\hat{W}(t) &= -\frac{1}{2\tau}[\hat{W}(t)\left[\frac{1}{1+2\epsilon\hat{L}^{-}\hat{L}^{\dagger}}-1\right]\hat{L}^{-},\hat{L}^{\dagger}] \\
&- \frac{1}{2\tau}\hat{\Lambda}[\hat{W}(t)\hat{L}^{-},\hat{L}^{+}], \qquad (18)
\end{aligned}$$

where

$$\hat{\Lambda}=\frac{1}{1+\epsilon[\ldots,\frac{1}{1+2\epsilon\hat{L}^{-}\hat{L}}\hat{L}^{-},\hat{L}^{+}]} \qquad (19)$$

is the operator which represents the expressions (16) and (17) by infinite series of small parameter, δ. We could formally represent the equation (17) compactly $[..., \frac{1}{1+2\epsilon\hat{L}^-\hat{L}}\hat{L}^-, \hat{L}^+]$ introducing the expression from the right hand side in each term of the decomposition of the Taylor series. It is observed that in the process of the Raman scattering the Bloch vector L^2 is conserved. In the next section we examine the solutions of this equation taking into consideration the higher orders in the above decomposition.

5. The Numerical Solution of the Master Equation

Using the decomposition on the small parameter of the system, we estimate in this section the quantum fluctuations as a function of photon number and the decomposition order of the small parameter, $x/(1+x)$, proposed in the last section. Let us start from the first order approximation, neglecting higher order terms in the decomposition (17)

$$\begin{aligned} \frac{d}{dt}\hat{W}(t) &= -\frac{1}{2\tau}[\hat{W}(t)\hat{D}^-\frac{1}{1+2\epsilon\hat{L}^-\hat{L}^\dagger}, \hat{L}^\dagger] \\ &+ H.c. \end{aligned} \tag{20}$$

The numerical solution of this equation is obtained, decomposing the density matrix on the angular momentum states, described by the eigan states of the operator $\hat{L}^2$, $\hat{W} = \sum_{m=-j}^{j} P_m(t)|j,m\rangle\langle m,j|$. Here the Hilbert vectors $|j,m\rangle$ belong to the three mode states in the resonator (Pump, Stokes and anti-Stokes), $P_m(t)$ is the population probability of the $|j,m\rangle$ state. Taking into consideration the action of the operators $\hat{L}^+$ and $\hat{L}^-$ on the angular momentum states, $\hat{L}^+|j,m\rangle = \sqrt{(j-m)(j+m+1)}|j,m+1\rangle$ and $\hat{L}^-|j,m\rangle = \sqrt{(j-m+1)(j+m)}|j,m-1\rangle$, from the master equation (20) we obtain the closed system of linear equations for the probabilities $\{P_m(t)\}$

$$\begin{aligned} \frac{d}{dt}P_m(t) &= -\frac{1}{\tau}\left\{\frac{(j-m)(j+m+1)}{1+2\epsilon(j-m-1)(j+m+2)}\right\}P_m(t) \\ &+ \frac{1}{\tau}\left\{\frac{(j-m+1)(j+m)}{1+2\epsilon(j-m)(j+m+1)}\right\}P_{m-1}(t). \end{aligned} \tag{21}$$

The second order approximation corresponds to the situation, in which the attenuation process of the photon conversion is significantly renormalized by the large number of the photon in the system, l:

$$\frac{d}{dt}\hat{W}(t) = -\frac{1}{2\tau}[\hat{W}(t)\hat{L}^{-}\frac{1}{1+2\epsilon\hat{L}^{-}\hat{L}^{\dagger}}, \hat{L}^{\dagger}] + \frac{1}{2\tau}\epsilon[[\hat{W}(t)\hat{L}^{-}, \hat{L}^{+}]\frac{1}{1+2\epsilon\hat{L}^{-}\hat{L}^{\dagger}}, \hat{L}^{+}] + H.c. \quad (22)$$

The numerical solution of this equation can be obtained after similar decomposition on the angular momentum states. In this case we can obtain instead of expression (21) the new equation which take into consideration the second order decomposition of (22). The new equation describes the attenuation of the conversion to the steady state on the Bloch-sphere surface

$$\begin{aligned} \frac{d}{dt}P_m(t) &= -\frac{1}{\tau}[\frac{(j-m)(j+m+1)}{1+2\epsilon(j-m-1)(j+m+2)} \\ &- \frac{\epsilon(j-m)^2(j+m+1)^2}{1+2\epsilon(j-m)(j+m+1)}]P_m(t) \\ &+ \frac{1}{\tau}[\frac{(j-m+1)(j+m)}{1+2\epsilon(j-m)(j+m+1)} - \frac{\epsilon(j-m+1)^2(j+m)^2}{1+2\epsilon(j-m+1)(j+m)} \\ &- \frac{\epsilon(j-m+1)(j-m)(j+m)(j+m+1)}{1+2\epsilon(j-m)(j+m+1)}]P_{m-1}(t) \\ &+ \frac{\epsilon}{\tau}[\frac{(j-m+2)(j-m+1)(j+m-1)(j+m)}{1+2\epsilon(j-m+1)(j+m)}]P_{m-2}(t), \\ & m = -j, -j+1, ..., j. \end{aligned} \quad (23)$$

We observe that with increasing of the order of the decomposition the number of the terms in the same order increases too. In this case it is better to take into consideration the next order of the decomposition on the small parameter, improving in this way the convergence of the right hand side of the equation (17)

$$\begin{aligned} &\frac{d}{dt}\hat{W}(t) = -\frac{1}{2\tau}[\hat{W}(t)\hat{L}^{-}\frac{1}{1+2\epsilon\hat{L}^{-}\hat{L}^{\dagger}}, \hat{L}^{\dagger}] \\ &+\frac{1}{2\tau}\epsilon[\{W(t)\hat{L}^{-}\hat{L}^{+} - \hat{L}^{+}W(t)\hat{L}^{-}\}\frac{1}{1+2\epsilon\hat{L}^{-}\hat{L}^{\dagger}}\hat{L}^{-}, \hat{L}^{+}] \\ &-\frac{1}{2\tau}\epsilon^2[\{W(t)\hat{L}^{-}\hat{L}^{+} - \hat{L}^{+}W(t)\hat{L}^{-}\}\frac{1}{1+2\epsilon\hat{L}^{-}L^{\dagger}}\hat{L}^{-}\hat{L}^{+}\frac{1}{1+2\epsilon\hat{L}^{-}\hat{L}^{\dagger}}\hat{L}^{-} \\ &-\hat{D}^{+}\{W(t)\hat{L}^{-}\hat{L}^{+} - \hat{L}^{+}W(t)\hat{L}^{-}\}\frac{1}{1+2\epsilon\hat{L}^{-}\hat{L}^{\dagger}}\hat{L}^{-}\frac{1}{1+2\epsilon\hat{L}^{-}\hat{L}^{\dagger}}\hat{L}^{-}, \hat{L}^{+}] \\ &+H.c. \end{aligned}$$

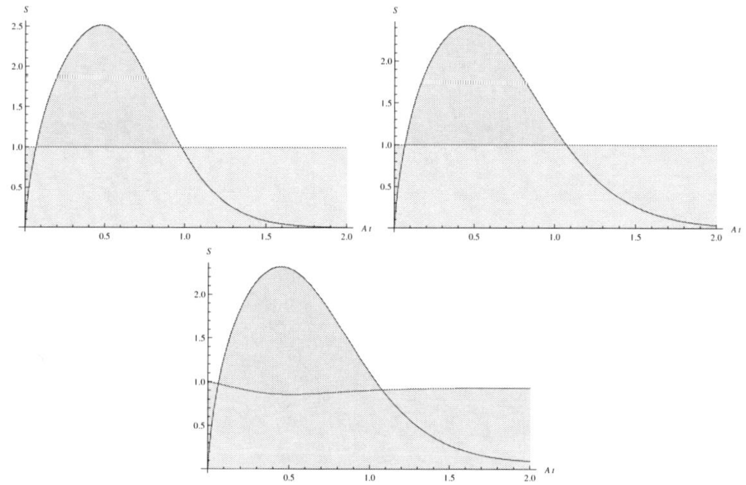

Figure 3. Comparison of Von Neumann entropy Von Neumann for first and second order on the decomposition on the small parameter for following parameters of the system $j = 6$, $\epsilon = 0.015$ calculated by the expressions (21), (23) and (24).

In the same decomposition order, we obtain the system of $2j + 1$ equations for the statistical weights P_m, which in comparison to right hand side of the system of equations (23) contains the corrections, $\Delta P(m)$, which are proportional to the ϵ^2. Let us rename the statistical probabilities in this approximation by the set of the functions Q_m. The system of equation for these statistical weights contains the corrections

$$\frac{d}{dt}Q_m(t) = Idem(Q_m) + \Delta Q(m)/\tau, \quad m = -j, ...j. \tag{24}$$

Here the $Idem(Q_m)$ is similar to the terms of the right hand side of the system of equations (23), in which the set of functions P_m we must replaced by Q_m.

Introducing this correction in the decomposition $\Delta Q(m)$,

$$\Delta Q(m) = -\frac{\epsilon^2(j-m)^3(j+m+1)^3}{[1+2\epsilon(j-m)(j+m+1)]^2}Q_m(t) + [\frac{\epsilon^2(j-m+1)^3(j+m)^3}{[1+2\epsilon(j-m-1)(j+m)]^2}$$
$$+\frac{\epsilon^2(j-m+2)^2(j-m+1)(j+m-1)^2(j+m)}{[1+2\epsilon(j-m+2)(j+m-1)][1+2\epsilon(j-m+1)(j+m)]}$$
$$+\frac{\epsilon^2(j-m+1)(j-m)^2(j+m)(j+m+1)^2}{[1+2\epsilon(j-m)(j+m+1)]^2}]Q_{m-1}(t)$$
$$-[\frac{\epsilon^2(j-m+2)^2(j-m+1)(j+m-1)^2(j+m)}{[1+2\epsilon(j-m+2)(j+m-1)][1+2\epsilon(j-m+1)(j+m)]}$$
$$+\frac{\epsilon^2(j-m+2)(j-m+1)^2(j+m-1)(j+m)^2}{[1+2\epsilon(j-m+1)(j+m)]^2}$$
$$+\frac{\epsilon^2(j-m+2)(j-m+1)(j-m)(j+m-1)(j+m)(j+m+1)}{[1+2\epsilon(j-m+1)(j+m)][1+2\epsilon(j-m)(j+m+1)]}]Q_{m-2}(t)$$
$$+\frac{\epsilon^2(j-m+3)[j^2-(m-2)^2][j^2-(m-1)^2](j+m)}{[1+2\epsilon(j-m+2)(j+m-1)][1+2\epsilon(j-m+1)(j+m)]}Q_{m-3}(t), \qquad (25)$$

we obtain the new system of equations (24). It observes the similar behavior: that the number of terms in the second order of perturbation decomposition (24) increases in comparison with the number of terms in the first order (23). As the statistical weights defined above satisfy the normalization conditions: $\sum_m P_m = 1$ and $\sum_m Q_m = 1$,we can introduce the factor $\Phi = \sum_m \sqrt{P_m Q_m}$, which help us to estimate the quality of the decomposition on the small parameter of the system. The decomposition becomes "good" if this factor achieves the value 1. In opposite case, when the factor, Φ, deviates from the unity, we must take into consideration the next decomposition on the small parameter.

The Fig. 3 represents the von Neumann entropy of the system, obtained from diagonalized representation through the statistical probabilities $S_t = \sum_{m=-j}^{j} P_m \log[P_m]$. We observe that with the increasing the order of decompositions the quantum fluctuations in the system increase too.

6. Three Particle Cooperative Emission in Single- and Two-Photon Exchanges Through Vacuum Field

Let us consider the interaction of three subsystems of radiators A, R, S, and D through the vacuum of electromagnetic field (EMF). The first two groups of radiators, R and S, prepared in excited state $|e_r\rangle \otimes |e_s\rangle$ can pass into Dicke super-radiance regime [17] relatively the dipole active transitions $e_a \rightarrow g_a$, $e_r \rightarrow g_r$ and $e_s \rightarrow g_s$ at frequencies ω_a, ω_r and ω_s (Figures 2 A , B). According to the Fig. 2 A the excited D atom relatively the dipole forbidden transition $2_d \rightarrow 1_d$ passes in the ground-state $|1_d\rangle$ simultaneously generating two quanta under the influences of first two atomic subsystems and vacuum fluctuation of EMF. This transition takes place through the virtual levels, which we replaced by the state $|3_d\rangle$ with opposite symmetry relative the ground $|1_d\rangle$ and excited $|2_d\rangle$ states. An example of two-photon spontaneous decay may be the transition relative the states with the same symmetry $|n+1)S\rangle \rightarrow |nS\rangle$ through the virtual state $|(n+1)P\rangle$ (see [25, 26]).

The new possibilities of cooperative emissions can be observed in the resonance interaction between the dipole active radiators of R, S subsystems and dipole forbidden of D subsystem in two situations: A. The emission frequencies of the dipole- active and dipole-forbidden radiators satisfy the resonance condition $\omega_r + \omega_s = \omega_d$. Here ω_r, ω_s are the transition frequencies of the R, S dipole-active radiators and $\hbar\omega_d = 2\hbar\omega_0$ is the energy distance between the ground $|nS\rangle$ and excited $|(n+1)S\rangle$ states of the dipole forbidden transitions of D radiator (see Fig.2A). The dipole active radiators A and S satisfy the scattering condition $\omega_a - \omega_s = \omega_d$ in the interaction with D subsystem. As it is represented in the Fig. 2 B, the transition energy of the dipole active atom A is larger than the transition energy between the levels of atoms S and D. Here, the two-photon resonance between the dipole-active subsystems A and S is more effective, when the intermediate state of the atom D has the small de-tuning from resonance $\Delta = |(E_3 - E_2)/\hbar - \omega_s|$.

Below we discuss the condition for which, the pure super-fluoresce of the small number of radiators [27]–[28], in the subsystems R, S and D enter into interaction during the delay time of cooperative spontaneous emission of each subsystem so that inhomogeneous broadening of excited atomic states can be neglected, $\tau_{r,i} \ll T_{2,i}$. Here $\tau_{r,i} = \tau_0/N_i$ is the collective time for which the polarization of the i-subsystem become macroscopic, $T_{2,i}$ is the de-phasing time of the subsystem i, which include the reciprocal inhomogeneous and Doppler-broadened line-width, $i \equiv S$, R, and D (see, for example, the papers [28, 29]). These conditions can be achieved using laser cooling method [30], [31] for three atomic ensembles represented in Figs. 2 A and B. Let us suppose, that delay time of the super-radiant pulse is less than $T_{2,i}$, we will drop the terms connected with de-phasing time $T_{2,i}$ from the kinetic equations.

In order to estimate the three particle interaction we will examine the situation in which the one and two-quantum interactions with the EMF bath are taken into account simultaneously. In this case it is necessary to eliminate from the density matrix equation the boson operators from the single and two photon interaction the EMF operator combinations. Let us consider the equations for the density matrix of the total system of radiators in interaction with the vacuum of EMF. In the interaction picture this equation is

$$i\hbar\frac{\partial\breve{\rho}(t)}{\partial t}=\lambda\big[\breve{H}_{I}(t),\breve{\rho}(t)\big],$$

where $\breve{H}_I(t)=\exp i\widehat{H}_o t/\hbar\widehat{H}_I\exp -i\widehat{H}_o t/\hbar$. Here the free part of the Hamiltonian is represented through the atomic and field operators

$$\breve{H}_0=\sum_k\hbar\omega_k\breve{a}_k^{\dagger}\breve{a}_k+\sum_{j=1}^{N_r}\hbar\omega_{jr}\breve{R}_{zj}+\sum_{l=1}^{N_a}\hbar\omega_{ls}\breve{A}_{zl}+\sum_{l=1}^{N_s}\hbar\omega_{ls}\breve{S}_{zl}+2\hbar\sum_{m=1}^{N}\omega_0\breve{D}_{zm},$$

Taking into consideration the conservation energy laws: $\hbar(\omega_r+\omega_s)=2\hbar\omega_0$

and $\hbar(\omega_r-\omega_s)=2\hbar\omega_0$ (according to Figure 2 A and B respectively), one can propose the following interaction Hamiltonian of the atomic system with EMF $\hat{H}_I=\hat{H}_{I1}+\hat{H}_{I2}$, where

$$\begin{aligned}\breve{H}_{I1} &= -\sum_k\sum_{n=1}^{N_a}(\mathbf{d}_a,\mathbf{g}_k)\breve{A}_j^{+}\breve{a}_k\exp\big[i(\mathbf{k},\mathbf{r}_j)\big]\\ &- \sum_k\sum_{j=1}^{N_r}(\mathbf{d}_r,\mathbf{g}_k)\breve{R}_j^{+}\breve{a}_k\exp\big[i(\mathbf{k},\mathbf{r}_j)\big],\\ &- \sum_k\sum_{l=1}^{N_s}(\mathbf{d}_s,\mathbf{g}_k)\breve{S}_l^{+}\breve{a}_k\exp[i(\mathbf{k},\mathbf{r}_l)]+H.c.; \end{aligned}\tag{26}$$

$$\begin{aligned}\breve{H}_{I2}=- &\sum_{k_1,k_2}\sum_{m=1}^{N}q_b(k_1,k_2)\breve{D}_m^{+}\breve{a}_{k_2}\breve{a}_{k_1}\exp[i(\mathbf{k}_1+\mathbf{k}_2,\mathbf{r}_m)]+H.c.\\ + &\sum_{k_1,k_2}\sum_{m=1}^{N}q_s(\mathbf{k}_1,\mathbf{k}_2)\breve{D}_m^{-}\breve{a}_{k_2}^{\dagger}\breve{a}_{k_1}(1-\delta_{k_1,k_2})\\ \times &\exp\big[i(\mathbf{k}_1-\mathbf{k}_2,\mathbf{r}_m)\big]+H.c.\end{aligned}\tag{27}$$

Here we have the first-order $\mathbf{g}_k = \sqrt{2\pi\hbar\omega_k/V}\,\epsilon_\lambda$, and second order

$$q_b(\mathbf{k}_1,\mathbf{k}_2) = \frac{(\mathbf{d}_{31},\mathbf{g}_{k_1})(\mathbf{d}_{32},\mathbf{g}_{k_2})}{2\hbar(\omega_{32}+\omega_{k_1})} + \frac{(\mathbf{d}_{31},\mathbf{g}_{k_2})(\mathbf{d}_{32},\mathbf{g}_{k_1})}{2\hbar(\omega_{31}-\omega_{k_2})},$$
$$q_s(\mathbf{k}_1,\mathbf{k}_2) = \frac{(\mathbf{d}_{31},\mathbf{g}_{k_1})(\mathbf{d}_{32},\mathbf{g}_{k_2})}{\hbar(\omega_{32}-\omega_{k_1})} + \frac{(\mathbf{d}_{31},\mathbf{g}_{k_2})(\mathbf{d}_{32},\mathbf{g}_{k_1})}{\hbar(\omega_{31}+\omega_{k_1})}$$

interaction constants. $\breve{a}_k$ and $\breve{a}_k^\dagger$ are annihilation and creation operators of EMF photons with wave vector $\mathbf{k}$, polarization ϵ_λ and frequency ω_k; $\mathbf{d}_a$, $\mathbf{d}_r$ and $\mathbf{d}_s$ are dipole momentum transitions between the ground and excited states for A, R and S atomic subsystems; d_{32} and d_{31} are dipole momentum transitions in the three-level system of the atomic sub-ensemble D. The excitation and lowering operators of A, R, S, and D atomic subsystems satisfies the commutation relations for $SU(2)$ algebra $[\breve{J}_l^+, \breve{J}_j^-] = 2J_{lz}\delta_{l,j}$; $[\breve{J}_{lz}, \breve{J}_j^\pm] = \pm\breve{J}_l^\pm\delta_{l,j}$, where $\breve{J}_l^\pm$ is equivalent with operators of $\hat{A}^\pm$, $\breve{R}_l^\pm$, $\breve{S}_l^\pm$ or $\breve{D}_l^\pm$ atomic subsystems. The inversion operator $\breve{J}_z$ is one of the $\hat{A}_z$, $\breve{R}_z$, $\breve{S}_z$ and D_z operators, respectively. The operators of EMF satisfy the commutation relation $[\breve{a}_k, \breve{a}_{k'}^\dagger] = \delta_{k,k'}$; $[\breve{a}_k^\dagger, \breve{a}_{k'}^\dagger] = 0$, where $k = (\mathbf{k},\lambda)$ is the wave vector and polarization of the photon.

In comparison with the paper [33, 32], here we will take into consideration simultaneously both effects connected with the influence of dipole active atoms A , S and A on the two-photon spontaneous emission of dipole forbidden D subsystem. Another attractive effect corresponds to the situation when the dipole active subsystems A, S and R are non-equidistant so that Dicke cooperative effects between each subsystem are negligible. Considering that exist the large number of possibilities between the subsystems S and A and subsystems S and R so that it is satisfied the two-photon resonance cooperative conditions $\omega_d = \omega_{a,l} - \omega_{s,m}$ and $\omega_d = \omega_{s,l} + \omega_{r,m}$ for different atomic pairs A_l and S_m ; R_m and S_l respectively. In this case three particle cooperative effects become significant and dipole active atoms drastically can modify the two-photon transition in the subsystem D.

Let us $\mathcal{P}$ is the projection operator for the complete density matrix $\breve{\rho}(t)$ on the vector basis of a free EMF subsystem $\rho_s(t) = \mathcal{P}\breve{\rho}(t)$, and $\breve{\rho}_b(t) = \bar{\mathcal{P}}\breve{\rho}(t)$, where $\breve{\rho}_s(t)$ and $\breve{\rho}_b(t)$ are slower and rapidly oscillating parts of the density matrix, respectively, $\overline{\mathcal{P}} = 1 - \mathcal{P}$. It can be shown that $\mathcal{P}^2 = \mathcal{P}$ and $\overline{\mathcal{P}}\mathcal{P} = 0$. Recognizing that for $t = 0$ atomic subsystem does not interact with EMF, we define the projection operator $\mathcal{P} = \breve{\rho}_{ph} \otimes Tr_{ph}\{\cdots\}$, where the trace it is taking over the EMF states, and $\breve{\rho}_0 = \rho_e(0)\breve{\rho}_{ph}(0)$, where $\breve{\rho}_{ph}(0) = |0\rangle\langle 0|$, represent the density matrix of the vacuum of EMF and $\rho_e(0)$ is the non-equilibrium distribution of the excitation on the atomic states represented in Fig. 2. In this case we represent the slow part of density matrix through the density matrix $\breve{W}(t) = Tr_r\{\breve{\rho}(t)\}$ of the atomic subsystem $\breve{\rho}_s(t) = \breve{\rho}_0 \otimes \breve{W}(t)$, where $\breve{W}(0) = Tr_{ph}\breve{\rho}(0) = \breve{\rho}_e(0)$. The equations for the matrix $\breve{\rho}_s(t)$ and $\breve{\rho}_b(t)$ are

$$\frac{\partial \breve{\rho}_s(t)}{\partial t} = -i\lambda \mathcal{P} L_I(t)\{\breve{\rho}_s(t) + \breve{\rho}_b(t)\}, \tag{28}$$

$$\frac{\partial \breve{\rho}_b(t)}{\partial t} = -i\lambda \overline{\mathcal{P}} L_I(t)\{\breve{\rho}_s(t) + \breve{\rho}_b(t)\}, \tag{29}$$

where $\lambda L_I(t) = [\breve{H}_I(t), \ldots]/\hbar$ is the interaction part of Liouville operator. Following the known procedure of elimination of the rapidly oscillating part of the density matrix, we integrate the equation (29) with respect to the $\breve{\rho}_b(t)$ and substitute the resulting solution in equation (28)

$$\frac{\partial \breve{\rho}_s(t)}{\partial t} = -\lambda^2 P \int_0^t d\tau L_i(t) U(t, t-\tau) L_i(t-\tau)\rho_s(t-\tau), \tag{30}$$

where

$$\breve{U}(t, t-\tau) = T \exp\Big\{ - i\lambda \overline{\mathcal{P}} \int_{t-\tau}^{t} d\tau_1 L_i(\tau) \Big\}.$$

The quantum correlation between the single and two-photon interaction of atoms with the thermal EMF can be found in the third order of the expansion on the small parameter λ on the right hand side of equation (30). Confining ourselves to the second and third order of the expansion on the small parameter λ, we represent the evolution operators $\breve{U}(t, t-\tau)$ and $\breve{\rho}_s(t-\tau)$ in the following approximate form $\breve{U}(t, t-\tau) \approx 1 - i\lambda \overline{\mathcal{P}} \int_{t-\tau}^{t} d\tau_1 L_i(\tau_1)$, and

$$\breve{\rho}_s(t-\tau) = \breve{\rho}_s(t) + \lambda^2 \mathcal{P} \int_0^{\tau} d\tau_1 L_i(t-\tau_1) \int_0^{t-\tau_1} d\tau_2 L_i(t-\tau_1-\tau_2)\breve{\rho}_s(t-\tau_1-\tau_2).$$

Upon substitution of this expression in equation (30), in the third order of small parameter λ the equation for $\rho_s(t)$ becomes

$$\frac{\partial}{\partial t}\breve{\rho}_s(t) = -\lambda^2 \mathcal{P} \int_0^t d\tau_1 L_i(t)\{L_i(t-\tau_1) - i\lambda \int_{t-\tau_1}^{t} d\tau_2 L_i(\tau_2)\} L_i(t-\tau_1)\breve{\rho}_s(t). \tag{31}$$

Let us represent the Liouville operator through single- and two-photon interaction parts with the thermal bath

$$L_i(t) = [\check{H}_{I1}(t), \ldots]/\hbar + [\check{H}_{I2}(t), \ldots]/\hbar.$$

Let us now study the third order contribution to the right hand side of the equation (31) Taking into account that the trace of an odd number of boson operator is zero, $Tr_{ph}\ \rho_0 \check{a}^\dagger_{k_1} \check{a}^\dagger_{k_2} \check{a}_{k_3} \check{a}_{k_4} \check{a}_{k_5}\} = 0$, it is not difficult to observe, that the projection of the operator product $\mathcal{P}\check{H}_{I1}\check{H}_{I2}\check{H}_{I2}$ takes a zero value too. In the third order of the small parameter λ the contribution of Liouville operator $\widehat{L}_{I1}$ and $\widehat{L}_{I2}$ must be found from the terms like $\mathcal{P}\widehat{L}_{I1}\widehat{L}_{I2}\widehat{L}_{I1}\widehat{\rho}_s(t)$, where $\lambda L^{s-}_{I2}(t) = [\check{H}^{s-}_{I2}(t), \ldots]/\hbar$ is the Liouville part for two-photon scattering process of D atomic subsystem and $\lambda L^{A-}_{I1}(t) = [\check{H}^{A-}_{I1}(t), \ldots]/\hbar$ corresponds to the single photon transition in A atom described by the Hamiltonian part (26) and (27), respectively. The scattering resonance can be represented by a diagram, in which we must take into consideration the conservation law $\omega_a - \omega_s = 2\omega_0$. We can find these diagrams of interaction between the A, S and D atoms in the following way. Considering that the trace of the odd number of boson operator is zero, $Tr_{ph}\{\check{\rho}_0 \check{a}^\dagger_{k_1} \check{a}_{k_2} \check{a}^\dagger_{k_3}\} = 0$, one can observe that the projection of the operator product $\mathcal{P}\check{H}_{I1}\check{H}_{I2}\mathcal{P} = 0$. Following this procedure of calculation of mean value of boson operators, it is observed that only the projections $\mathcal{P}\check{H}_{I1}\check{H}_{I1}\mathcal{P} \neq 0$, $\mathcal{P}\check{H}_{I2}\check{H}_{I2}\mathcal{P} \neq 0$ and $\mathcal{P}\check{H}_{I1}\check{H}_{I1}\check{H}_{I2}\mathcal{P} \neq 0$ give the main contribution. So that after the trace on the EMF variables we obtain $Tr\{\rho_{ph}\check{a}_{k_1}\check{a}_{k_3}\check{a}^\dagger_{k_2}\check{a}^\dagger_{k_4}\} = [\delta_{k_1 k_2}\delta_{k_3,k_4} + \delta_{k_1,k_4}\delta_{k_3,k_2}]$, $Tr\{\rho_{ph}\check{a}_{k_1}\check{a}^\dagger_{k_2}\} = \delta_{k_1 k_2}$, $Tr\{\rho_{ph}\check{a}^\dagger_{k_2}\check{a}^\dagger_{k_4}\check{a}_{k_1}\check{a}_{k_3}\} = 0$ we can find the correlations between A, R, S and D atomic subsystem represented in the Fig.2. Following this procedure, we observe that according to the representation of the exchange integrals between the radiators the trace of the right hand side of the expression (31) can be regarded as the sum of three terms

$$\frac{d\check{W}(t)}{dt} = \frac{d\check{W}^{(0)}(t)}{dt} + \frac{d\check{W}^{(21b)}(t)}{dt} + \frac{d\check{W}^{(21s)}(t)}{dt}, \tag{32}$$

The first term describes the cooperative single and two-photon effects in each subsystem respectively. The second term describes the exchanges between single photon processes of R ,S subsystems and two-photon transitions of the D radiators as this is represented in Fig.2A. The third term describes the three particle scattering effect represented in Fig.2B. All parameters and collective exchange integrals between the radiators of each S, R and D subsystems are defined in the literature [17, 34]

$$\frac{d\check{W}^{(0)}(t)}{dt} = \frac{i}{\hbar}[\check{H}_0, \check{W}(t)] + \sum_{\alpha=a,s,r} \frac{1}{2\tau_\alpha} \sum_{l,j=1}^{N_\alpha} \chi_\alpha(j,l)[\check{J}^-_{j\alpha}, \check{W}(t)\check{J}^+_{l\alpha}]$$
$$+ \frac{1}{2\tau_d} \sum_{l,j=1}^{N} \chi_d(j,l)[\check{D}^-_j, \check{W}(t)\check{D}^+_l] + H.c., \tag{33}$$

where $\tau_\alpha = 3\hbar c^3/(4d_\alpha^2\omega_\alpha^3)$ is the spontaneous emission time of the dipole active α-atom, $\tau_d = \pi 3^2\hbar^2 c^6/(4^2\omega_0^7 d_{23}^2 d_{31}^2 q_b^2(\omega_0,\omega_0))$ is two photon spontaneous emission rate. This equation can be used for description of interaction between the dipole forbidden and dipole active subsystems of radiators. We give here for comparison the real part of the single-photon exchange integrals $Re[\chi_\alpha(j,l)] = \sin(\omega_\alpha r_{j,l}/c)/[\omega_\alpha r_{j,l}/c]$ of A, S and R dipole active atomic subsystems. The real part of two-photon exchange integral is proportional to $\sin^2(\omega_0 r_{j,l}/c)/[\omega_0 r_{j,l}/c]^2$.

Let us now found the three particle exchanges between the radiators, represented in Figure 2 A and described by the second term of the master equation (32)

$$\frac{d\check{W}^{(21b)}(t)}{dt} = -\frac{i}{2\tau^b_{srd}} \sum_{m=1}^{N} \sum_{j=1}^{N_r} \sum_{l=0}^{N_s} \Big\{ U_b(j,m,l)[\check{D}^-_m, \check{W}(t)\check{R}^+_j \check{S}^+_l]$$
$$+ U_b^*(j,l,m)[\check{S}^+_l, [\check{R}^+_j, \check{D}^-_m \check{W}(t)]] \Big\}$$
$$- \frac{i}{2\tau^b_{srd}} \sum_{m=1}^{N} \sum_{j=1}^{N_r} \sum_{l=0}^{N_s} V_b(j,m,l) \Big\{ [\check{D}^-_m \check{W}(t)\check{S}^+_l, \check{R}^+_j(t)]$$
$$+ [\check{D}^-_m \check{W}(t)\check{R}^+_j, \check{S}^+_l] \Big\} + H.c. \tag{34}$$

Here for $\omega_s \simeq \omega_r$, we have founded the following integrals

$$\frac{1}{\tau^b_{srd}} = \left\{\frac{4}{3}\right\}^2 \frac{\omega_s^3(\omega_r)^3 d_s d_r d_{23} d_{31}}{2\hbar^2 c^6} \left[\frac{1}{\omega_{32}+\omega_s} + \frac{1}{\omega_{31}+\omega_r}\right],$$
$$V_b(j,m,l) \simeq \frac{c^2[\exp[-i\omega_s r_{ml}/c] - 1][\exp[i\omega_r r_{jm}/c] - 1]}{\omega_s\omega_r r_{jm} r_{ml}},$$
$$U_b(j,m,l) = \exp[-i\omega_s r_{ml}/c] V^b(j,l,m). \tag{35}$$

Here $1/\tau^b_{srd}$ is the cooperative emission time of three atoms belonging to S, R and D subsystems situated at a relatively small distance $r_{jl} \ll \lambda_{s(r)}$; $V^b(j,m,l)$ is the exchange integral, which describe the influence of the m atom from D subsystem on the single photon transitions of the j and l radiators from R and S subsystems. $U_b(j,m,l)$ is the inverse influence of j and l radiators from R and S subsystems on the two-photon transitions of m radiator from the D subsystem.

In the case, $r_{jl} \ll \lambda_{s(r)}$, for three atoms, we can introduce the collective operators $\check{C}^+ = \check{S}^+\check{R}^+$, $\check{C}^- = \check{S}^-\check{R}^-$ and $\check{C}_z = (\check{S}_z+\check{R}_z)/2$, which satisfy the commutation relations for $su(2)$ algebra $[\check{C}_z, \check{C}^+] = \check{C}^+$, and $[\check{C}^+, \check{C}^-] = 2\check{C}_z$. It is interesting that the new operators commute with the difference $\check{V}_z = (\check{S}_z - \check{R}_z)/2$. Here we observe the analogy between cooperative operators introduced in this section and collective photon operators $\hat{L}^+$, $\hat{L}^-$ and $\hat{L}_z$ described in the Sec. 3

Let us now found the part of the master equation (32) for resonance scattering interaction between the absorbed and emitted photons by the dipole active subsystems S and A and D. In this situation, the scattering part of the master equation for the case represented in Fig. 2 can be represented by the expression

$$\begin{aligned}
\frac{d\check{W}^{(21s)}(t)}{dt} &= -\frac{i}{\tau^s_{sad}}\sum_{m=1}^{N}\sum_{j=1}^{N_a}\sum_{l=0}^{N_s} V_s(j,m,l)[\check{D}^-_m(t), \check{S}^-_l\check{W}(t)\check{A}^+{}_j] \\
&+ \frac{i}{2\tau^s_{sad}}\sum_{m,j,l=1}\Big\{U_s(j,m,l)[\check{A}^+_j, \check{S}^-_l\check{D}^-_m\hat{W}(t)] \\
&+ U^*_s(j,m,l)[\check{S}^-_l, \check{W}(t)\check{A}^+{}_j\check{D}^-_m]\Big\} + H.c., \qquad (36)
\end{aligned}$$

where

$$\begin{aligned}
\frac{1}{\tau^s_{sad}} &= \left(\frac{4}{3}\right)^2\frac{d_sd_ad_{23}d_{31}\omega_s^3(\omega_a)^3}{c^6\hbar^2}\left[\frac{1}{\omega_{32}-\omega_s}+\frac{1}{\omega_{31}+\omega_s}\right], \\
V_s(j,m.l) &= \frac{(\exp[-i\omega_s r_{ml}/c]-1)(\exp[i\omega_a r_{mj}/c]-1)}{\omega_s\omega_a(r_{ml}/c)(r_{mj}/c)}, \\
U_s(j,m,l) &= -\frac{c^2[\exp[i\omega_a r_{jm}/c]-1][\exp[i\omega_s r_{ml}/c]-1]}{\omega_a\omega_s r_{mj}r_{ml}}. \qquad (37)
\end{aligned}$$

The first term in the equation (36) describes the transition of D-atom under the influence of the scattering process of emitted photons by the atoms from A and

S subsystems. This process is described by exchange integral $V_s(j, m.l)$. The last two terms in the master equation (36) describe the scattering process of emitted photons by the atoms S and A under the influence of D subsystem, and are described by three particle exchange integral $U_s(j, m, l)$.

For the system of three radiators, situated at the distance less emission wavelength, we observe that $V_s(j, m, l) \approx 1$ and $U_s^*(j, m, l) \approx 1$. In this case it is possible to introduce the new collective operators for atom A and S: $\breve{F}_z = (\breve{A}_z - \breve{S}_z)/2$ $\breve{F}^- = \breve{S}^+\breve{A}^-$ and $\breve{F}^+ = \breve{S}^-\breve{A}^+$ These operators satisfy the commutation relations for the $SU(2)$ group too $[\breve{F}^+, \breve{F}^-] = 2\breve{F}_z$, $[\breve{F}_z, \breve{F}^\pm] = \pm\breve{F}^\pm$. As in the case of two photon interaction the operator $\breve{U}_z = (\breve{A}_z + \breve{S}_z)/2$ commutes with the lowering and exciting operators, $\breve{F}^-$ and $\breve{F}^+$.

7. Cooperative Migration of Excitations between Three Radiators in Scattering Process

The master equation (32) can be used for description of cooperative interaction between the dipole forbidden and dipole active subsystems of radiators in two-photon excitation. Indeed, passing again from Schrodinger to Heisenberg pictures $Tr[\hat{W}(t)\hat{O}(0)] = Tr[\hat{W}(0)\hat{O}(t)]$, from the equations (32) we can obtain the equation of the arbitrary atomic operator. Let us discuss the four particle interaction in which the S-atom enters both resonances represented in Fig. 2 A and B. The closed system of equations for correlation functions can be found studding the chain of questions for these operators of the dipole forbidden D subsystem and dipole active A, S and R radiators. Considering, that the numbers of atoms in the subsystems are relatively small, the chain of equations is obtained from the kinetic equations for the inversion operators of each subsystem. According to the master equation (32 this coin of the equation can be reduced to closed systems of equations for the small numbers of atoms. Indeed, we can introduce the mean value of the exited numbers of atoms of type A, S, R, and D atoms: $\langle\widehat{N}_\alpha\rangle = \langle\widehat{J}_{z\alpha}(t)\rangle + 0.5$, $\alpha = a, s, r, d$. Here the operator $\widehat{J}_{z\alpha}$ is considered equivalent to the invasions of each atom $\widehat{A}_z$, $\widehat{S}_z$, $\widehat{R}_z$ and $\widehat{D}_z$. Let we considered the simplest situation, when three dipole active atoms A, R and S are situated at distance r relatively the radiator D smaller than the radiation wavelengths. Introducing the correlation function between the polarizations of three atoms $\langle F_b\rangle = i[\langle\widehat{D}^+(t)\widehat{S}^-(t)\widehat{R}^-(t)\rangle - \langle\widehat{S}^+(t)\widehat{R}^+(t)\widehat{D}^-(t)\rangle]$, and $\langle F_s\rangle = i\{\langle \widehat{A}^+(t)\widehat{D}^-(t)\widehat{S}^-(t)\rangle - \langle \widehat{S}^+(t)\widehat{D}^+(t)\widehat{A}^-(t)\rangle\}$ from the master equation (32) we obtain the closed system of equations for these atomic correlations

$$\begin{aligned}
\frac{d}{dt}\langle \widehat{N}_s\rangle &= -\frac{\langle \widehat{N}_s\rangle}{\tau_s} - \frac{1}{2\tau^b_{srd}}\langle \widehat{F}_b\rangle - \frac{1}{2\tau^s_{sad}}\langle \widehat{F}_s\rangle,\\
\frac{d}{dt}\langle \widehat{N}_r\rangle &= -\frac{\langle \widehat{N}_r\rangle}{\tau_r} - \frac{1}{2\tau^b_{srd}(x)}\langle F_b\rangle,\\
\frac{d}{dt}\langle \widehat{N}_a\rangle &= \frac{\langle \widehat{N}_r\rangle}{\tau_a} + \frac{1}{2\tau^s_{sad}}\langle \widehat{F}_s\rangle,\\
\frac{d}{dt}\langle \widehat{N}_d\rangle &= -\frac{\langle \widehat{N}_d\rangle}{\tau_d} - \frac{1}{2\tau^b_{srd}(x)}\langle \widehat{F}_b\rangle + \frac{1}{2\tau^s_{sad}(x)}\langle \widehat{F}_s\rangle,
\end{aligned}$$

$$\begin{aligned}
\frac{d}{dt}\langle \widehat{F}_b\rangle &= -\frac{A_b}{2}\langle F_b\rangle\rangle + \frac{2}{\tau^b_{srd}}\Big[5\langle \widehat{N}_s\widehat{N}_r\widehat{N}_d\rangle\\
&- \langle \widehat{N}_s\widehat{N}_r\rangle/2 - 3\langle \widehat{N}_s\widehat{N}_d\rangle/2\\
&- 3\langle \widehat{N}_r\widehat{N}_d\rangle/2 + \langle \widehat{N}_d\rangle\Big]\\
\frac{d}{dt}\langle \widehat{F}_s\rangle &= -\frac{2}{\tau^s_{srd}}\Big\{\big[\langle \widehat{N}_s\widehat{N_a}\widehat{N}_d\rangle - \langle \widehat{N}_s\widehat{N_a}\rangle/2\big]\\
&+ \Big[2\langle \widehat{N}_s\widehat{N_a}\widehat{N}_d\rangle - \langle \widehat{N}_a\widehat{N}_s\rangle - 0.5\langle \widehat{N}_s N_d\rangle\\
&+ 0.5\langle \widehat{N}_a\rangle - 0.5\langle \widehat{N}_a N_d\rangle\Big]\Big\} - \frac{A_s}{2}\langle \widehat{F}_s(t)\rangle,\\
\frac{d}{dt}\langle \widehat{N}_s\widehat{N}_r\widehat{N}_d\rangle &= -A_b\langle \widehat{N}_s\widehat{N}_r\widehat{N}_d\rangle,\\
\frac{d}{dt}\langle \widehat{N}_s\widehat{N}_r\rangle &= -B_b\langle \widehat{N}_s\widehat{N}_r\rangle + \frac{1}{2\tau^b_{srd}}\langle F_b\rangle,\\
\frac{d}{dt}\langle \widehat{N}_r\widehat{N}_d\rangle &= -D_b\langle \widehat{N}_s\widehat{N}_r\rangle,\\
\frac{d}{dt}\langle \widehat{N}_s\widehat{N_a}\widehat{N}_d\rangle &= -A_s\langle \widehat{N}_s\widehat{N_a}\widehat{N}_d\rangle,\\
\frac{d}{dt}\langle \widehat{N}_s\widehat{N}_a\rangle &= -B_s\langle \widehat{N}_s\widehat{N}_a\rangle,\\
\frac{d}{dt}\langle \widehat{N}_s\widehat{N}_d\rangle &= -C\langle \widehat{N}_s\widehat{N}_d\rangle - \frac{1}{2\tau^s_{srd}}\langle \widehat{F}_s\rangle,\\
\frac{d}{dt}\langle \widehat{N}_a\widehat{N}_d\rangle &= -D_s\langle \widehat{N}_a\widehat{N}_d\rangle. \qquad (38)
\end{aligned}$$

Here, the following correlation function between the atomic excitations it is introduced: for the three cooperative processes represented in the Fig.2A: $\langle \widehat{N}_s \widehat{N}_r \widehat{N}_d \rangle$, $\langle \widehat{N}_s \widehat{N}_r \rangle$, $\langle \widehat{N}_s \widehat{N}_d \rangle$,$\langle \widehat{N}_r \widehat{N}_d \rangle$ and Fig.2B: $\langle \widehat{N}_s \widehat{N}_a \widehat{N}_d \rangle$, $\langle \widehat{N}_s \widehat{N}_a \rangle$, $\langle \widehat{N}_a \widehat{N}_d \rangle$. In the system of equation (38) it is defined the following rates: $A_b = 1/\tau_r + 1/\tau_d + 1/\tau_s$, $B_b = 1/\tau_r + 1/\tau_s$, $C = 1/\tau_d + 1/\tau_s$, $D_b = 1/\tau_d + 1/\tau_r$, $A_s = 1/\tau_a + 1/\tau_d + 1/\tau_s$, $B_s = 1/\tau_a + 1/\tau_s$, $D_b = 1/\tau_d + 1/\tau_a$.

Let us now study the transferring of excitation from the S radiator to the R-atom. For simplicity, we consider that the A atom is absent from the system. In Fig. 4 A, we have plotted the excitation of the R atom from the ground state, when the atoms S and D are in an excited state. We observe that the spontaneous generation of photons by the atoms D and S can effectively exit the dipole active atom R, when the dipole active radiator S is in the excited state. In the opposite case, when the atom S is in the ground state the excitation of the atom R is nonsignificant. The similar mutual three particle cooperative excitations appear when the atom D is prepared in the ground state and the atoms S and R are in the excited states. This export of the excitation from the radiators R and S to the excited state of the dipole forbidden radiator D is represented in Fig. 4B.

The possibility of non-resonant excitation transfer between the atoms A and S represented in Fig. 2B can be found in the special preparation of the system. The effective energy transfer of the excitation from atom A to atom S through the ground state of the dipole forbidden transitions of D-radiator is represented in the Fig. 5A. In opposite situation, when the atom D is in the excited state, this energy transfer from the A atom to S atom is less then in first case (see see Fig. 5B).

As follows from the system (38), the first $\langle \widehat{N}_d \rangle / \tau_d$ and the second terms $(1/\tau^s_{sad}) \langle \widehat{F}_s \rangle$ describe the generation rate of entangled photon pairs and scattering rate with absorption of Stokes photon and the generation of anti-Stokes, respectively. When the time tends to infinity, all excited atomic energies $E_0 = \hbar\omega_a + \hbar\omega_s + \hbar\omega_d$ of atoms S, A and D are emitted by the system. Taking into account the conservation law in the process two-photon emission $\omega_{k_1} + \omega_{k_2} - \omega_d = 0$ and the scattering process $\omega_a - \omega_s - \omega_d = 0$, we observe, that the last process becomes predominant, when the cooperative scattering rate $1/\tau^s_{sad}(x)$ increase. In other words, the probability of absorption of Stokes photon $\hbar\omega_s$ and generation the new anti-Stokes photon $\hbar\omega_a$ by D-atom becomes possible. In this case three atoms represented in Figure 2 B can generate two anti-Stokes photons with energy $E_0 = 2\hbar\omega_a$.

It was proposed the projection operator method which describes the effec-

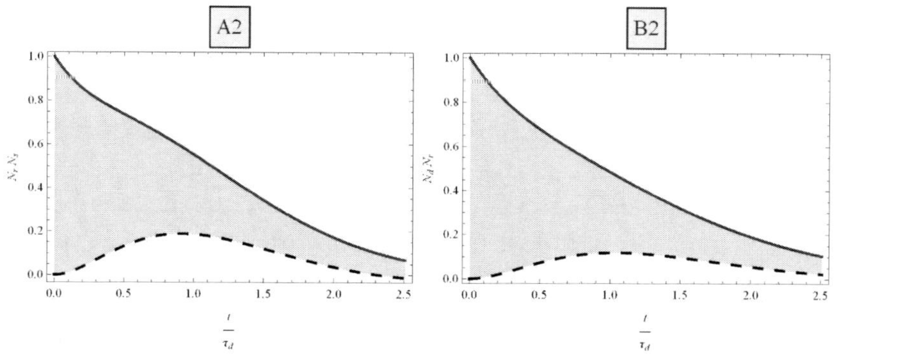

Figure 4. A. The excitation of R radiator (dashed line) by the decay process dipole active S-atom (thick line). The process takes place through the exited D atom for following parameters of the system : $\langle N_r \rangle = 0$, $\langle N_s \rangle = \langle N_d \rangle = 1$ $\tau_d/\tau_r = \tau_d/\tau_s = 1$, $0.5\tau_d/\tau^b_{srd} = 0.8$, $\tau_d/\tau^s_{sad} = 0$. B. The excitation of dipole forbidden D radiator (dashed line) by the decay process dipole active of S and R atoms (thick line) following parameters of the system : $\langle N_d \rangle = 0$, $\langle N_s \rangle = \langle N_r \rangle = 1$. Other parameters are as in Fig.A2.

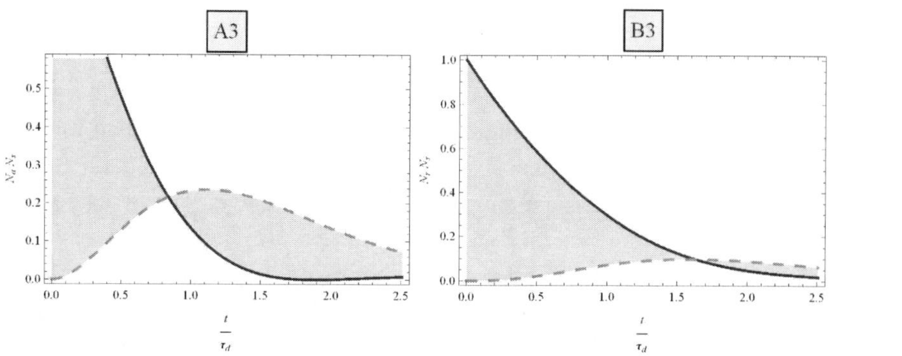

Figure 5. A. The excitation of S radiator (dashed line) by the decay process dipole active A-atom (thick line). The process takes place through the ground state of D atom for following parameters of the system : $\langle N_a \rangle = 1$, $\langle N_s \rangle = \langle N_d \rangle = 0$. $\tau_d/\tau_a = \tau_d/\tau_s = 1$, $0.5\tau_d/\tau^s_{sad} = 1.$, $\tau_d/\tau^b_{srd} = 0$. B. The excitation of dipole active S radiator (dashed line) by the decay process of A and D radiators. Here it is used the following parameters of the system: $\langle N_d \rangle = \langle N_a \rangle = 1$, $\langle N_s \rangle = 0$. Other parameters are as in Fig. A.

tive interaction between three subsystems of radiators in two-photon resonance. The new cooperative migration energy from one excited dipole active atom to another is proposed in the non-resonant case. This excitation transfer takes place through dipole forbidden atom prepared in the excited or ground state as this is represented in Fig 4. It is studied two situations. In the first case the S-atom is in an excited state and R-atom in the ground. The second situation corresponds to scattering transfer of this energy between the excited state of A radiator and ground state of A atom. This transfer is more effective when the dipole forbidden atom is prepared in the ground state as this is represented in Fig. 5A. When the atom D is in the excited state, the emitted Stokes photon by an atom of the system S can be absorbed by the radiator D so that three radiators pass into the ground state generating two anti-Stokes photons with energies $E_0 = 2\hbar\omega_r$. The opposite situation can be observed when D atom is prepared in the ground-state. In this case, the absorption of the anti-Stokes photon generated by A-atom is accompanied by generation of another Stokes photon by D atom according to Figure 2B.

The recent experiments about two-photon coincidences focus on the studies of the decay of heavy atomic systems. While some predictions of theory are available for the hydrogen-like atoms or ions, the promising experiment connected with *Rb* atom placed in the single-mode cavity in bi-photon resonance with a dipole forbidden transition $40S \rightarrow 39S$ in cascade configuration [26] opens the new possibilities for the realization of collective effects between three radiators. The modern super-fluorescent experiments with cold atomic systems in the cavity [30, 31] give us larger possibilities in the experimental estimations of relative rates between single- and two-photon decay processes τ_d/τ_{rsd}^b and $\tau_d/\tau_{s(r)}$. The ionizing method of atomic flux passing through the cavity can be used in these experimental testing migrations of the excitation from one atom to another in three particle resonance proposed in this section. This transfer between two quantum oscillators off from resonance take place through the dipole forbidden transition of thread atom as this is represented in Fig. 2.

8. Coherent Properties of Two-mode Fields and Their Application in Quantum Holography and Lithography

Let us consider two nonlinear processes of light generation in lasers and collective decay phenomena. In the second order of interaction of light with matter these processes strongly connect two waves in the output detection region and give us the possibility to study the coherent effects between these fields. One of these effects is connected with lithographic optical limits in measurements. Taking into account the concept about the dropped lithographic limit $\Delta \geq \lambda/2$ in two quanta coherent processes, A. N. Boto et al. [35] have observed the new lithographic limits in parametric down conversion effect. In this process it in generated two waves with frequencies ω_s- signal andω_i-idler. The lithographic limit for these two waves is smaller than traditional limit $\Delta \geq \lambda/4$. Following this conception, we introduced the similar coherence between the photons in Raman lasing processes. In this case one photon from driving field is absorbed (Stokes photon a_s) and other photon is generated (anti-Stokes b_a^+). Taking into account the product of Sakes and anti-Stokes amplitudes$P^-(t,z)$, one can construct the common phase interference figures between two space points z and $z+\Delta$. These interfere figures are similar with coherence coincidences observed in double slits experiment with single amplitude coherent field, $G_2(\Delta) = \langle P^-(z)\, P^+(z+\Delta)\rangle$.

A. Let us firstly describe the non-linear effects in which coherence between two photons is possible. The experiments for the determination of two-photon decay time for metastable states for hydrogen-like or helium-like atoms (see Fig. 6) relatively to dipole-forbidden transitions $2^2S_{1/2} \rightarrow 1^2S_{1/2}$ and $2^1S_0 \rightarrow 1^1S_0$ was studied in the experiments [36].

The two-photon cooperative phenomena can appear in similar experiments if at the distance less than the mean value of radiation wavelength $\lambda_m = 2hc/(E_{2S} - E_{1S}) = 2400[A^0]/Z^2$ certain number of inverted atoms $N \approx 10 - 100$ is situated. In this case the transition probability is found in the second order of perturbation theory $M(k_1,k_2)\delta(\hbar\omega_{k_1} + \hbar\omega_{k_2} - (\mathrm{E}_{2s} - E_{1s}))$ of the living time of metastable state $2S$. The cooperative decay time, $\tau_r = \tau_0/N$, delay time of super-radiant pulse $t_0 = (\tau_0/N)lnN$, decreases in accord with the super-radiance low. Here $\tau_0 = 0.121 * 10^{-6}$ is the two-photon decay time for hydrogen- or helium-like atoms, Z is the order number of the element. In

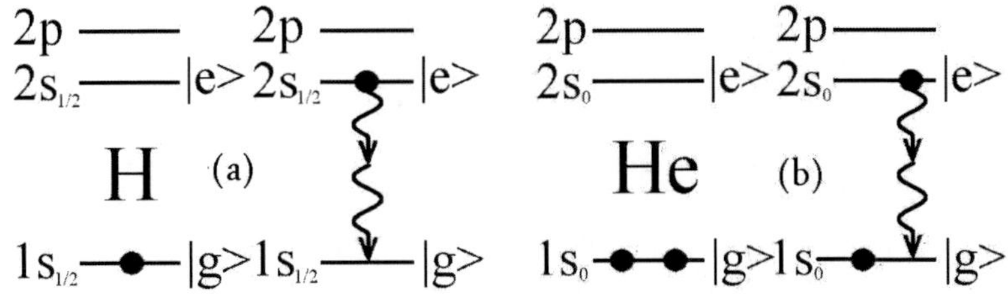

Figure 6. The two-photon spontaneous emission in Hydrogen-like (a) and Helium-like (b) atoms.

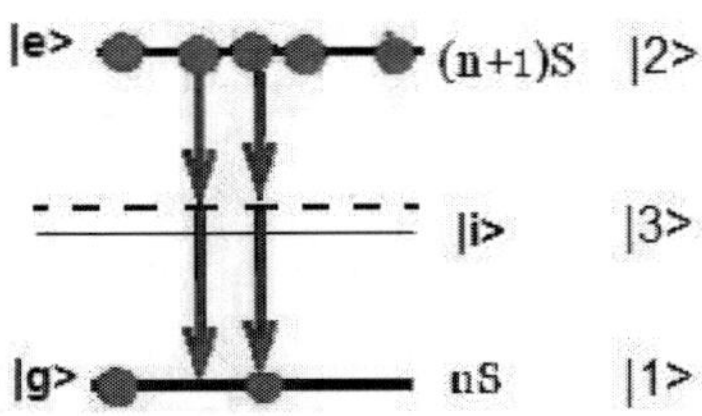

Figure 7. Two-photon cooperative decay stimulated by discrete mode structure of micro-cavity (chain line) observed in experiment [26].

extended systems of atomic radiators the super-fluorescence is observed in the case when the mean distance between the atoms is of the order of the radiation wavelength [18]. These experiments give the hope that in gamma diapason, when the mean distance between the inverted nuclei is comparable with the radiation wavelength, the super-radiance can be observed too. An interesting effect of two-photon cooperative emission is possible in micro-cavities. In this case the mode structure of the cavity stimulates the two-photon decay effect in comparison with cascade effect [37, 26] (see Fig. 7)

The property of entanglement between the photons emitted in the two photon lasers and the parametrical dawn conversion has a great Impact on application in quantum information. The possibility of induced two-photon generation per atomic transition was observed by Nikolaus, Zhang and Toschek [36].

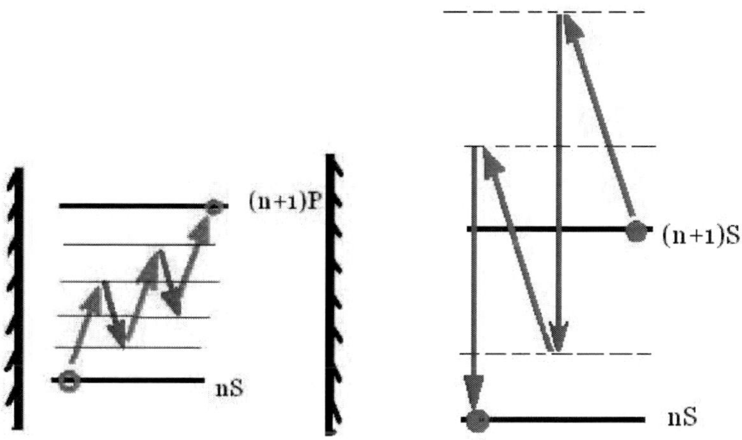

Figure 8. Sources of entangled photons in multi-photon processes. (a) scattering diagram for two-photon laser beam amplification in laser-driven two-level atoms; (b) hyper Raman four-photon process with emission of two big energy entangled photons and absorption of two photons with smaller energy.

The scattering effects in two-photon amplifier attenuate the possibility to realize two-photon lasing.

The first experiments demonstrated that two-photon amplification and lasing in the presence of external sources are possible [36, 36] (see Fig. 8). These ideas open the new conception about the coherence. Indeed, introducing the amplitude of two-boson field, generated in two-photon lasers, we can observe that the generation amplitude is described by a product of field amplitudes

$$P^{(+)}\left(t,z\right) = E_b^{(+)}(z,t)E_a^{(+)}(z,t) = G(k_b,k_a)ab\exp[i\omega_0 t - i(k_a+k_b)z]; \quad (39)$$

where $\omega_0 = \omega_b + \omega_a = \omega_{21}$ is the total frequency of generated photons. In this case we can introduce the following operators of bi-bosin field

$$I^+ = b^+a^+;\ I^- = ab;\ \ I_z = (a^+a + b^+b)/2. \quad (40)$$

In with coherent state of single photon lasers we can introduce the following coherent state for these bi-boson fields

$$|\mu\rangle = \exp\{\mu I^+\}|j,j.\rangle\{1-|\mu|^2\}^j, \quad (41)$$

which belong to $SU(1,1)$ symmetry represented in the Figs. 6 and 8. Here μ is the coherent displacement of bi-bosin oscillator. According to this conception the function $P^{(+)}(t,z)$ has the same behavior as the electrical component of single photon laser. For example, the mean value of this function on the coherent state (41) can be represented through the harmonic functions with phase and amplitude

$$\left\langle P^{(+)}(t,z)\right\rangle = P_0 \exp[i\omega_0 t - i(k_a + k_b)z + \varphi], \tag{42}$$

where $P_0 = G(k_b, k_a)\left|\left\langle \hat{a}\hat{b}\right\rangle\right|$ is the amplitude and $\varphi = Arg\left\langle \hat{a}\hat{b}\right\rangle$ is the phase of electrical field strengths of two fields *a* and *b*. The Fig. 9a represents the time or z- dependence of such functions. The lithographic limit follows from the difference between the maximum and minimum of two slits experiment represented in the Fig.9b. According with expression (42) and the distinguish distance Δ between the slits follows that the second order correlation function $G_2(\Delta) = \langle P^-(z)\, P^+(z+\Delta)\rangle$ pass from maximal to minimal values for $\Delta(k_1 + k_2)\sin\theta = \pi$. From this expression follows that lithographic limit is realized for $\sin\theta \to 1$. Or the distance between two stresses on the (x,y) plan mast be $\Delta \to \lambda_a\lambda_b/[2(\lambda_a + \lambda_b)]$. In the case when the wavelengths of the fields "*a*" and "*a*" coincide $\lambda = \lambda_a = \lambda_b$ we obtain the lithographic limit proposed in paper [35]

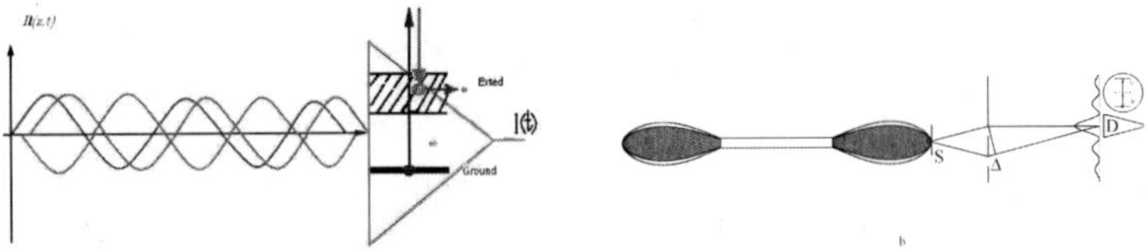

Figure 9. (a) The properties of two-ray coherence have similar behaviour as a coherent strength of single photon laser field. (b) The double-slit experiment with interference between pairs of modes which form the strength product of EMF. The interference picture can be observed in photon excitation of the detector.

B. Let us propose the coherent phenomena, which can appear between two fields in Raman processes. If we study generation of Stokes light under the non-coherent pumping with anti-Stokes field, we can introduce the following representation of the bi-boson field

$$\Pi^-(t,z) \;-\; E_s^{(+)}(z,t)E_a^-(z,t) = G(k_s,k_a)\hat{a}_s\hat{b}_a^+\exp[i\omega_0 t - i(k_a - k_s)z], \quad (43)$$

where $E_s^{(+)}(z,t)$ and $E_a^-(z,t)$ are positive and negative defined strength of Stokes and anti-Stokes field (see Fig. 5a), $\hat{a}_s^+$, $\hat{b}_a^+$and $\hat{a}_s$, $\hat{b}_a$ are the annihilation and creation operators of the electromagnetic field at Stokes, ω_s, and anti-Stokes, ω_a, frequencies respectively; $\omega_a - \omega_s = \omega_0$ is the fixed frequency of the bi-bozon field accordingly with the transition diagram represented in Fig.(10)a.

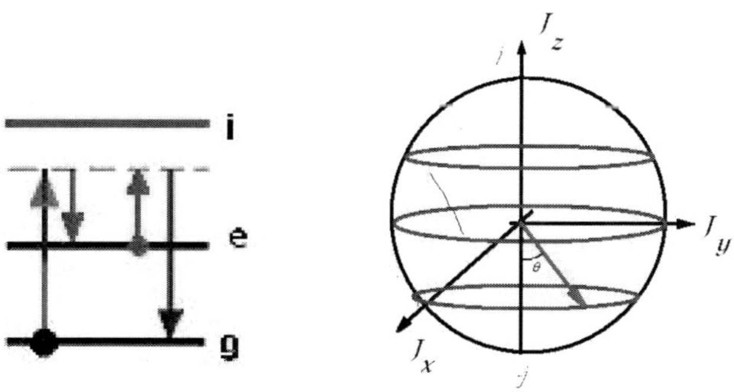

Figure 10. Scattering generation diagram with the generation of Stokes (or anti-Stokes) photons and the representation of the symmetry of such generation in phase space of bimodal field.

According to this definition one can introduce the new bi-boson operators $\hat{J}^- = \hat{b}^\dagger\hat{a}$; $\hat{J}^+ = \hat{a}^+\hat{b}$; $\hat{J}_z = (\hat{a}^+\hat{a} - \hat{b}^+\hat{b})/2$ which for a constant number of photons in resonator it is possible the conservation of Kasimir vector, $J^2 = \hat{J}_z^2 + \hat{J}_x^2 + \hat{J}_y^2$ represented in Fig.(10) (here $\hat{J}_x = (\hat{J}^+ + \hat{J}^-)/2$, $\hat{J}_y = (\hat{J}^+ - \hat{J}^-)/2i)$). If we consider that at initial time the number of anti-Stokes photons is N=2j, one describes the two photon scattering lasing processes by coherent state for this bi-boson field belonging to SU(40) algebra

$$|\alpha\rangle = \exp\{\alpha\hat{J}^+\}|-j,j\rangle\{1+|\alpha|^2\}^{-j}, \quad (44)$$

where $\alpha = \tan(\theta/2)$ is the amplitude of this bi-boson field obtained in the Raman lasing processes. Taking into account the coherent state 44, one can found the mean value of strength product $\left\langle\hat{\Pi}(t,z)\right\rangle = [\left\langle\hat{\Pi}^-(t,z)\right\rangle + \left\langle\hat{\Pi}^+(t,z)\right\rangle]/2$

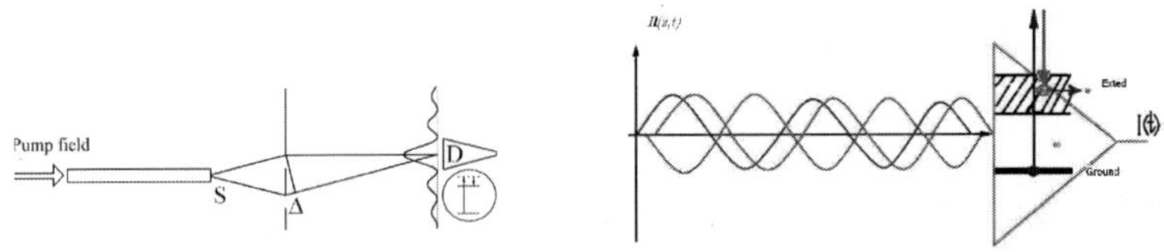

Figure 11. The pump of bimodal field in the Raman scattering effect in pencil shape cavity according to the scheme represented in Fig. 9. The time dependence of the mean value of vector $\langle P(z,t)\rangle$ and method of detection.

$$\left\langle \hat{\Pi}\left(t,z\right)\right\rangle \;=\; \Pi_0\cos[\omega_0 t-(k_a-k_s)z+\varphi], \tag{45}$$

where $\Pi_0 = G(k_s,k_a)\left|\left\langle \hat{a}_s\hat{b}_a^+\right\rangle\right|$ and $\varphi = \arg\left\langle \hat{a}_s\hat{b}_a^+\right\rangle$ are the amplitude and phase of bimodal field formed from Stokes and anti-Stokes photons. The minimal distance between maximal and minimal values of amplitude of correlation function $G_2\left(\Delta\right) = \left\langle \hat{\Pi}^-\left(z\right)\hat{\Pi}^+\left(z+\Delta\right)\right\rangle$ in the double-slit experiments observed with two-photon detectors corresponds to the lithographic limit of this conjugate entangled bimodal field (see Fig. 11)

This distance is inverse proportional with wave vectors difference $\Delta \geq \pi/(k_a - k_s)$ and of course is manifested in the second order coherence between the Stokes and anti-Stokes fields. We emphasize here that Stokes and anti-Stokes field can be incoherent field in its boson representation. Of course, these two field representations must be regarded as a bi-boson coherent state of light formed from the action of the operator product of the Stokes and anti-Stokes fields $\hat{a}_s\hat{b}_a^+$ on the vacuum of electromagnetic field respectively. If the frequency of anti-Stokes field is larger than Stokes frequency, we can approximate this limit with the expression $\Delta \geq \lambda_a/2$, λ_a and λ_a are the anti-Stokes and Stokes wave length respectively. This coherent phenomenon between Stokes and anti-Stokes fields can be used in Holography and other coherent processes with phase memory.

The minimal distance between maximal values of amplitude of above correlation corresponds to lithographic limits of this conjugate entangled bimodal field, $\Pi^-\left(z\right)$. This distance is inverse proportional with wave vector difference $\Delta \geq \pi/(k_a - k_s)$ and of course is manifested in the second order coherence between the Stokes and anti-Stokes modes. We emphasize here that

Stokes and anti-Stokes modes of EMF can be incoherent field in its boson representation. Of course these two field representation must be regarded as a bi-boson coherent state of light formed from the action of operator product of Stokes and anti-Stokes fields $a_s b_a^+$on the vacuum of EMF respectively: $|\mu\rangle = \exp[\mu \hat{a}_s \hat{b}_a^\dagger - \mu \hat{a}_s^\dagger \hat{b}_a]|0\rangle$, where μ is the amplitude of this bi-boson field, obtained in the Raman lasing processes, $|0\rangle$ is the vacuum state of this bi-boson field. If the frequency of anti-Stokes field is larger than Stokes frequency, we can approximate this limit with the expression $\Delta \geq \lambda_a/2$, $\lambda_a$and λ_a are the anti-Stokes and Stokes wave length respectively. This coherent phenomenon between Stokes and anti-Stokes modes can be used in Holography and other coherent processes with phase memory. Below we discuss how hologram can be constructed using the recording phase information of a bimodal field in the optical sensitive medium to this phase, using two separate beams of bimodal field (one is the "usual" beam associated with the image to be recorded and the other is a known as the reference beam). Exporting the interference pattern between the bi-bason modes, described above in principle this becomes possible. For example the Stokes and anti-Stokes fields can be regarded as a field formed from the electromagnetic strength product (43), so that the common phases $\phi = \omega_0 t - k_0 z$ of this two field amplitude $\hat{a}_s$ and $\hat{b}_a^\dagger$ has similar behaviour as the phase of single mode coherent field, where $\omega_0 = \omega_a - \omega_s$ and $k_0 = k_a - k_s$ are the frequencies and wave vectors difference between the Stokes and anti-Stokes modes, respectively. The coherent propriety of this product of the electric field components is proposed to apply in the possibilities to construct the time, space holograms of the real objects, taking into account the conservation of the phase of the amplitude product in the propagating and Interference processes.

Today it is proposed the lot of experiments in which is manifested holographic principals of processing of quantum information. One of them is the model of prof. Teich with co-authors [13]. According with this model the correlation between the entangled photons obtained in parametric down conversion can be used in quantum holography.

In Fig. 12 the principles of such holography is given. Here S is a source of entangled-photon pairs. C is a (remote) single photon-sensitive integrating sphere that comprises the wall of the chamber concealing the hidden object (bust of Plato). D is a (local) $2 - D$ single-photon-sensitive scanning or array detector. $h1$ and $h2$ represent the optical systems that deliver the entangled photons from S to C and D, respectively. The quantity $p2\ x2$ is the marginal

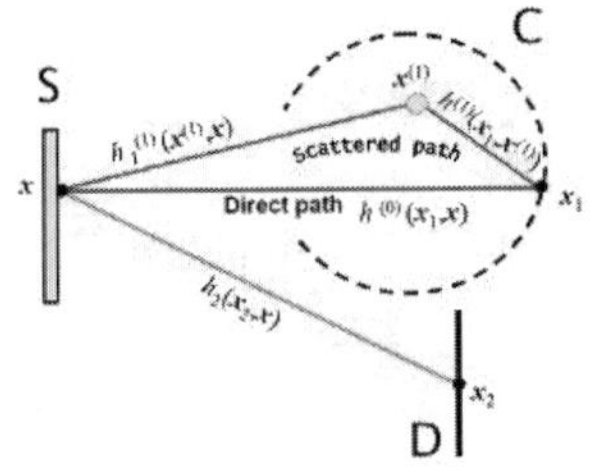

Figure 12. The hologram principle of registration of information with entangled photons proposed in paper [13].

coincidence rate, which is the hologram of the concealed object. Thin and thick lines represent optical and electrical signals, respectively. The hologram in the parametric down conversion it is realized in terms of the correlations between the entangled photon in the single pair. The coherence between the pairs is not taken into consideration.

Following the idea of second order coherence proposed in papers [18, 35], we propose the new type of hologram registration in bimodal photon field based on the coherent properties (42)- (45). As well known the holographic code in single photon coherent effects is based on mixing the original wave (hereafter called the object wave) I_0 with a known reference wave I_r and recording their interference pattern in the $z = 0$ plane. The intensity of the sum of the two waves is photographically recorded and a transparency of complex amplitude transmittance T, proportional to the intensity, is proposed in the literature (see for example [38]. Following this idea we can construct such a hologram using the vector (39) for two-photon coherence or vector (43) for scattering bimodal field. Indeed, according to this classical definition one can represent the original bimodal field through $P_O(z,t) = (P_O^+(z,t) + P_O^-(z,t))/\sqrt{2}$ and known reference bimodal wave through $P_r(z,t+\tau) = (P_r^+(z,t+\tau)+P_r^-(z,t+\tau))/\sqrt{2}$. In this case the scheme of interference pattern is represented in Fig. 13 and has many analogies with classical holograms. The transmittance is given by interference between the original and reference bimodal wave in the point in point $z = 0$

$$T = \langle : (P_O(z,t) + P_r(z,t+\tau))^2 : \rangle = \langle P_O^-(0,t)P_O^+(0,t)\rangle + \langle P_r^-(0,t+\tau)P_r^+(0,t+\tau)\rangle$$
$$+\langle P_O^-(0,t)P_r^+(0,t+\tau)\rangle + \langle P_r^-(0,t+\tau)P_O^+(0,t)\rangle. \tag{46}$$

This transmission can be detected by two photon detectors on the plane (x, y) and the interpretation of imaging can be expressed in classical terms

$$T = G_{O2} + G_{r2} + 2\sqrt{G_{r2}G_{O2}}\cos[\arg(P_O^+(0,t)) - \arg(P_r^+(r,t+\tau))] \tag{47}$$

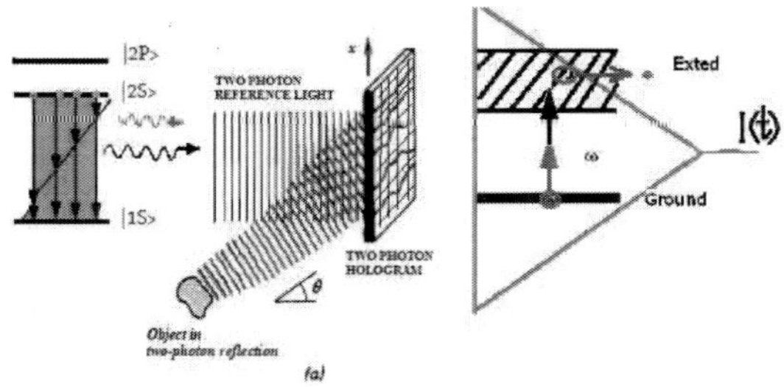

Figure 13. Two-photon coherent light and principle of hologram registration taking into consideration the phase and amplitude of bimodal field.

Here $G_{O2} = \langle P(0,t)P_U^+(0,t)\rangle$ is the intensity of the detected bi-photons from the original bimodal field; $G_{r2} = \langle P_r^-(0,t+\tau)P_r^+(0,t+\tau)\rangle$ is the intensity of detected bi-photons from reference bimodal wave from the object. The phase dependence of the imagine can be described by argument of complex number $\langle P_O^-(0,t)P_r^+(0,t+\tau)\rangle$ or $\arg(P_O^+(0,t)) - \arg(P_r^+(r,t+\tau))$. The same behaviour has the bimodal field formed from Stokes and anti-Stokes photons. In this case in expressions (46) and (47) the field $P^+(0,t)$ must b) must be substituted with $\Pi^+(0,t)$. The propriety of last bimodal field is described by the expressions (43)-(45). The detection scheme on the plan $z = 0$ is represented in Fig. 11.

In this section was proposed the new possibilities in decreasing of lithographic limit using coherent bimodal electromagnetic field formed from the Stokes and anti-Stokes waves. If initially the field is prepared in the Stokes state with photon number $N = 2j$ as this is proposed in the Ref.[11] and atomic stream in the excited state the lithographic limit is described by expression $\Delta \geq \lambda_a\lambda_a/(2(\lambda_s - \lambda_a))$ according to $SU(2)$ symmetry. In the case when the wavelength of generating anti-Stokes field belong to x-ray spectral region the lithographic limit can achieve the Nano-scale diapason. This is an important limit for modern nano-electronics and bio-physics.

The application of coherent effect of bimodal field in holography opens the new perspectives in the transmission of information not only through entangled state of photons but also through the second order coherence. This method of recording of information opens the new perspectives in quantum cryptography and quantum information. Today it is proposed the lot of experiments in which is manifested holographic principles of processing of quantum information (see for example [39, 40]. All these methods open new perspectives in the coding and decoding of information.

References

[1] Pan, J.-W. , Simon, C. , Brukner, C. & Zeilinger, A. (2001). Entanglement purification for quantum communication. *Nature* [doi:10.1038/35074041], 410, 1067-1070.

[2] Walther, P., Resch, K. J. and Brukner C. & Zeilinger, A. (2006). Experimental Entangled Entanglement. *Phys. Rev. Lett.* [doi = 10.1103/PhysRevLett.97.020501], 97(2), 020501-4.

[3] O'Brien, J. L., Furusawa, A. & Vuckovic, J. (2009). Photonic quantum technologies. *Nature the Photonics*[doi:10.1038/nphoton.2009.229], 3, 687 - 695.

[4] Banaszek, K., Demkowicz-Dobrzanski, R. & Walmsley, I. A. (2009). Quantum states made to measure. *Nature Photonics* [doi:10.1038/nphoton.2009.223], 3, 673 - 676.

[5] Horodecki, P. & Horodecki, R. (2001). Distilation and bound entanglement. *Quantum Information and Computation*, 1(1), 45-75.

[6] Acan A. (2009). Entanglement: Entangled and bound. *Nature Physics*, 5, 711 - 712.

[7] Amselem E. & Bourennane, M. (2009). Experimental four-qubit bound entanglement. *Nature Physics* 5, 748 - 752.

[8] Lavoie, J., Kaltenbaek, R., Piani, M. & Resch, K.J. (2010). Experimental bound entanglement in a four-photon state. *Phys. Rev. Lett.* [doi = 10.1103/PhysRevLett.105.130501] 105(13), 130501-4.

[9] Kaneda, F. ; Shimizu, R. ; Misumori, Y. ; Kosaka, H. & Edamatsu, K. (2011). Generation and caracterisation of bound entanglement in optical qbits. *Progress in Informatics (Special issue: Quantum information technology)*, 8, 27-31.

[10] Enaki, N. A. (1988). Superradiation from two-photon spontaneous decay. *Sov. Phys. JETP*], 67, 2033-2038; [(1988)*Zh. Eksp. Teor. Fiz.*, 94, 135]. Enaki N. , Macovei M. (1997). Cooperative emission in the process of cascade and dipole-forbidden transitions *Phys. Rev. A* , 56, 3274-3286.

[11] Enaki, N. & Turcan, M. (2012). Cooperative scattering effect between Stokes and anti-Stokes field stimulated by a stream of atoms. *Optics Communications* [doi: 10.1016/j.optcom.2011.11.011], 285(5), 686-692.

[12] Enaki, N. & Turcan, M. (2013). Cooperative quantum correlations between Stokes and anti-Stokes modes in four-wave mixing. *Physica Scripta* [doi:10.1088/0031-8949/2013/T153/014021] 153 014021.

[13] Abouraddy, A. F., Saleh, B. E. A. and Sergienko, A. V. & Teich, M. C. (2001). Role of Entanglement in Two-Photon Imaging. *Phys. Rev. Lett.* [doi = 10.1103/PhysRevLett.87.123602], 87(4), 123602-4.

[14] Glauber, R.J. (1963). Coherent and Incoherent States of the Radiation Field. *Phys. Rev.* [doi = 10.1103/PhysRev.131.2766], 131(6), 2766–2788.

[15] Sudarshan, E. C. G. (1963). Equivalence of Semiclassical and Quantum Mechanical Descriptions of Statistical Light Beams. *Phys.Rev. Lett.* [doi = 10.1103/PhysRevLett.10.277], 10(7), 277–279.

[16] Belinskii, A.V. & Klyshko, D.N. (1993). Interference of light and Bell's theorem. *Sov. Phys. Phys. Usp.* 36(8) 653?693.

[17] Dicke, R. H. (1954). Coherence in Spontaneous Radiation Processes. *Phys. Rev.*, 93, 99-110.

[18] Enaki, N. A. (1988). Superradiation from two-photon spontaneous decay. *Sov. Phys. JETP*], 67, 2033-2038; [(1988)*Zh. Eksp. Teor. Fiz.*, 94, 135].

[19] Perina, J. (1991). *Quantum Statistics of Linear and Non-linear Optical Phenomena Kluwer*, Dordrecht.

[20] Enaki, N.& Rosca, T. (2011). The exact quantum solution of N-radiators in cooperative interaction with a cavity field. Phys. Scr. [doi:10.1088/0031-8949/2012/T147/014011], T147, 014011(8pp)

[21] Gauthier D.J., Chapter 10, in *Two photon lasers*, Durham, USA, 2002.

[22] Radcliffe, J.M. (1971). J. Phys. A , 4, 313-323; Arecchi, J.F.T., Courtens, E. ; Gilmore, E. & Thomas, H. (1972). Phys. Rev. A, 6, 2211-2237.

[23] Enaki, N.A. (2010). Master equation and its Fokker-Planck representation in two quantum interactions of radiators with squeezed field J. Mod.Optics, 57, 1397-1407.

[24] Gauthier, D.J., Wu, Q., Morin, S.E. & Mossberg T.W. (1992). Realization of a continuous-wave, two-photon optical laser. *Phys.Rev. Lett.* [doi = 10.1103/PhysdoiRevLett.68.464], 68(4), 464-467.

[25] Marrus, R. & Schmieder, R. W. (1972). Forbidden Decays of Hydrogenlike and Heliumlike Argon *Phys. Rev. A*, 5(3), 1160–1173.

[26] Brune, M., Raimond, J. M. & Haroche, S. (1987). Theory of the Rydberg atom two-photon micromaser. *Phys. Rev. A* 35, 154-163; Brune , M., Raimond, J. M. , Goy, P., Davidovich, L. & Haroche S. (1987). Realization of a two-photon maser oscillator. *Phys. Rev. Lett.*, 59, 1899-1992.

[27] Bonifacio R & Lugiato L A (1975). Cooperative radiation processes in two-level systems: Superfluorescence *Phys. Rev. A.* 11 1507-1521.

[28] Lehmberg, R H (1970). Radiation from an N-Atom System. I. General Formalism. *Phys. Rev. A*, 2 883-888.

[29] Lugiato L A, Oldano C, Narducci L M (1988). Cooperative frequency locking and stationary spatial structures in lasers. *J. Opt. Soc. Am. B* 5, 879-888.

[30] Wang T., Yelin S. F., Ct R., Eyler E. E., Farooqi S. M., Gould P.L., Koďż žddtrun M., Tong D., Vrinceanu D. (2007) Superradiance in ultracold Rydberg gases *Phys. Rev. A* 75 033802(4 p.).

[31] Paradis E., Barrett B., Kumarakrishnan A., Zhang R. & Raithel G. (2008).Observation of superfluorescent emissions from laser-cooled atoms. *Phys. Rev. A*,77, 043419 (7 p.).

[32] Enaki,N. (2013). Three-particle cooperative emission in single- and two-photon exchanges through the vacuum field. *Physica Scripta* [doi:10.1088/0031-8949/2013/T153/014020], 153, 014020.

[33] Enaki N. A. (2012) Cooperative Resonance Interaction Between One-and Two-Photon Super-fluorescences Trough the Vacuum Field *J. of. Phys.: Conf. Ser.* [doi:10.1088/1742-6596/338/1/012005],338, 012005-12.

[34] Enaki, N.A. (2012). Mutual cooperative effects between single- and two-photon super-fluorescent processes through vacuum field. *Eur. Phys. J. D* [DOI: 10.1140/epjd/e2012-20563-1], 66(4):98.

[35] Boto, A. N. , Kok, P., Abrams, D. S., Braunstein, S. L., Williams, C. P. & Dowling, J. P. (2000). Quantum Interferometric Optical Lithography: Exploiting Entanglement to Beat the Diffraction Limit. *Phys. Rev. Lett.* [doi = 10.1103/PhysRevLett.85.2733], 85(13), 2733–2739.

[36] Nikolaus, B., Zhang, D. Z. & Toschek, P. E. (1981). Two-Photon Laser. *Phys. Rev. Lett.* [doi = 10.1103/PhysRevLett.47.171], 47(3), 171–173.

[36] Ali, R., Ahmad, I. , Dunford R. W., Gemmell, D. S., Jung, M., Kanter, Mokler, P. H., Berry, H. G., Livingston, A. E. , Cheng S. & Curtis, L. J. (1997). Shape of the two-photon-continuum emission from the 1s2s1S0 state in He-like krypton. *Phys.Rev. A*, 55, 994-1006..

[37] Enaki, N.A & Macovei, M.A. (1999). Two-photon cooperative decay in a cavity in the presence of a thermalized electromagnetic field. *Rus. Phys. JETP* [http://link.springer.com/article/10.1134/1.558839], 88, 633-641.

[38] Saneh, B. E. A. & Teich M.C. (1991). *Fundameltals of Photonics*, ISBNs: 0-471-83965-5 , (Hardback), 947p.

[39] Vasilyev, D.V., Sokolov, I.V. & Polzik, E.S. (2008). Quantum memory for images: A quantum hologram. *Phys. Rev. A* [doi = 10.1103/PhysRevA.77.020302], 77(2), 020302-4.

[40] G Bjrk, G, Sderholm, J. & Snchez-Soto, L. L. (2004). Two-photon imaging and quantum holography. *J. Opt. B: Quantum Semiclass*, [doi:10.1088/1464-4266/6/6/003], 6(6), S478-S482.

In: New Developments in Quantum Optics ... ISBN: 978-1-63483-394-3
Editor: Nicole Stewart, pp. 99-111

Chapter 3

POLARIZATION AND ENTANGLEMENT OF THREE-PHOTON BELL-DIAGONAL MIXED STATES

Iulia Ghiu*
Centre for Advanced Quantum Physics, Department of Physics,
University of Bucharest, Bucharest-Măgurele, Romania

Abstract

We investigate two quantities that characterize the two-mode states of the radiation field: quantum degree of polarization and entanglement. We evaluate analytically the quantum degree of polarization for the three-photon Bell-diagonal states by exploiting two distance-based measures: Hilbert-Schmidt and Bures. Two examples are considered further in our analysis: the Werner state and a family of Bell-diagonal states defined by a parameter. For each example we compare the concurrence with the two types of quantum degrees of polarization mentioned above.

PACS: 03.67.Mn, 03.65.Ca, 42.50.Dv

Keywords: Two-mode state, entanglement, quantum degree of polarization, Bell-diagonal states

*E-mail address: iulia.ghiu@g.unibuc.ro

1. Introduction

Analyzing different properties, such as entanglement and quantum degree of polarization, for two-mode states of the radiation field is of great importance for a deeper understanding of the correlations that occur in the system. Polarized states of the quantum electromagnetic field are used in many experiments of quantum information processing, e.g., Bell inequalities [1], quantum cryptography [2], [3], quantum teleportation [4], [5], superdense coding [6], entanglement swapping [7], entanglement purification for quantum communication [8], and quantum computation [9].

The degree of polarization is defined in classical optics with the help of the Stokes parameters [10]. The straightforward generalization to the quantum regime is made by replacing the Stokes parameters by the expectation values of the Stokes operators [11]. Since the photon polarization is a resource that can be widely used due to the possibility of its almost lossless manipulation, a proper quantification of polarization is needed. There are many proposals for defining the quantum degree of polarization. One possibility is to employ the distance between the density operator of the given field and the set of unpolarized states. Recently, two metrics have been used for defining the quantum degree of polarization: the Hilbert-Schmidt and Bures [12], [13], [14].

On the other hand, the key role of entanglement in quantum information theory is well-known [15]. An analysis of both the entanglement and polarization was done in the case of continuous variable systems for photon-added coherent states [16].

The purpose of this chapter is to present a comparison between entanglement and quantum degree of polarization for discrete systems, namely for two-mode three-photon states of the radiation field. We find that the two measures are incomparable, i.e. there exist states for which the quantum degree of polarization is greater than the entanglement and vice versa, i.e. states where the entanglement is greater than the degree of polarization.

The chapter is organized as follows. We start by reviewing the recently formulated requirements to be fulfilled by any acceptable measure of quantum degree of polarization in Sec. 2. We present further two distance-based measures that will be employed for evaluating the degree of polarization for a two-mode state. The states that are investigated in Sec. 3 are a particular set of two-mode mixed states of the quantum radiation field, namely the family of three-photon states which are Bell-diagonal. We consider the concurrence as a measure of entanglement in our analysis. The expressions of two measures of the degree of polarization: Hilbert-Schmidt and Bures are obtained for the Bell-diagonal states. Sec. 4 is devoted to two applications, namely the Werner state and a family of Bell-diagonal states that depends on a parameter. In each case, we plot two graphics that show the comparison between the concurrence and the quantum degree of polarization. Our conclusions are outlined in Sec. 5.

2. Preliminaries: Quantum Degrees of Polarization

We consider a mixed two-mode state $\hat{\rho}$ of the quantum radiation field. We use the notation $|\, k, N-k\,\rangle = |\, k\,\rangle_H \otimes |\, N-k\,\rangle_V$, with $k = 0, 1, ..., N$ for the standard two-mode Fock basis, where H means horizontal polarization, while V the vertical one. The N-photon states $|\, k, N-k\,\rangle$ span an SU(2) invariant subspace of dimension $N+1$.

Any polarization transformation is generated by the Stokes operators $\hat{S}_1$, $\hat{S}_2$, and $\hat{S}_3$:

$$\begin{aligned}
\hat{S}_1 &:= \hat{a}_H^\dagger \hat{a}_V + \hat{a}_H \hat{a}_V^\dagger, \\
\hat{S}_2 &:= \frac{1}{i}\left(\hat{a}_H^\dagger \hat{a}_V - \hat{a}_H \hat{a}_V^\dagger\right), \\
\hat{S}_3 &:= \hat{a}_H^\dagger \hat{a}_H - \hat{a}_V^\dagger \hat{a}_V,
\end{aligned}$$

that are built with the amplitude operators $\hat{a}_H$ and $\hat{a}_V$ of the horizontal and vertical modes, respectively.

The class of linear polarization transformations is a group of unitary operators $\hat{U}_{\text{pol}}$ on the two-mode Hilbert space $\mathcal{H}_H \otimes \mathcal{H}_V$. The operators $\hat{U}_{\text{pol}}$ form an infinite-dimensional unitary representation of the group SU(2), whose parametrization in terms of the Euler angles ϕ, θ, ψ is as follows:

$$\hat{U}_{\text{pol}}(\phi, \theta, \psi) = \exp\left(-i\,\frac{\phi}{2}\,\hat{S}_3\right)\exp\left(-i\,\frac{\theta}{2}\,\hat{S}_2\right)\exp\left(-i\,\frac{\psi}{2}\,\hat{S}_3\right). \quad (1)$$

A state $\hat{\sigma}$ that remains invariant under any polarization transformation (1), i.e. $\hat{U}_{\text{pol}}\hat{\sigma}\hat{U}_{\text{pol}}^\dagger = \hat{\sigma}$, is unpolarized [17], [18]. The spectral decomposition of this state $\hat{\sigma}$ is [17], [18], [19], [20], [21]

$$\hat{\sigma} = \sum_{N=0}^{\infty} \pi_N \frac{1}{N+1}\,\hat{P}_N, \quad (2)$$

where

$$\hat{P}_N := \sum_{n=0}^{N} |n, N-n\rangle\langle n, N-n|$$

is the projection operator onto the vector subspace of the N-photon states. The parameters π_N satisfy the normalization condition

$$\sum_{N=0}^{\infty} \pi_N = 1.$$

Let us consider the Fock expansion of an arbitrary two-mode state $\hat{\rho}$ of the radiation field:

$$\hat{\rho} = \sum_{M=0}^{\infty}\sum_{N=0}^{\infty}\sum_{m=0}^{M}\sum_{n=0}^{N} |m, M-m\rangle\langle m, M-m|\hat{\rho}|n, N-n\rangle\langle n, N-n|. \quad (3)$$

The expression (3) can be split into the sum of the block-diagonal terms i.e. $M = N$ and that of the off-block-diagonal ones that are obtained for $M \neq N$. We denote by $\hat{\rho}_b$ the block-diagonal density matrix associated with the given state $\hat{\rho}$ [14]:

$$\hat{\rho}_b := \sum_{N=0}^{\infty} p_N \hat{\rho}_N. \quad (4)$$

In Eq. (4), p_N is the probability of the Nth excitation manifold:

$$p_N = \mathrm{Tr}(\hat{\rho}\hat{P}_N) = \sum_{n=0}^{N} \rho_{nn}^{(N)}, \quad (5)$$

where

$$\rho_{mn}^{(N)} := \langle m, N-m|\hat{\rho}|n, N-n\rangle.$$

A von Neumann measurement of the number of photons gives the outcome N with the probability p_N, while the output state denoted by $\hat{\rho}_N$ is [14]:

$$\hat{\rho}_N := \frac{1}{p_N}\hat{P}_N\hat{\rho}\hat{P}_N = \frac{1}{p_N}\sum_{m=0}^{N}\sum_{n=0}^{N} \rho_{mn}^{(N)} |m, N-m\rangle\langle n, N-n|, \qquad p_N > 0.$$

A proper measure of the quantum degree of polarization, denoted by $\mathbb{P}$, has to fulfill three requirements [14]:

- $\mathbb{P}(\hat{\rho}) = 0$ if and only if $\hat{\rho}$ is unpolarized.
- $\mathbb{P}(\hat{\rho})$ is invariant under polarization transformations

$$\mathbb{P}(\hat{U}_{\text{pol}}\, \hat{\rho}\, \hat{U}^{\dagger}_{\text{pol}}) = \mathbb{P}(\hat{\rho}).$$

- $\mathbb{P}(\hat{\rho})$ is not affected by coherences between different excitation manifolds, i.e. it is defined with the help of the block-diagonal density matrix $\hat{\rho}_b$ given by Eq. (4)

$$\mathbb{P}(\hat{\rho}) := \mathbb{P}(\hat{\rho}_b). \tag{6}$$

In this chapter we make a comparison between the concurrence and two types of quantum degrees of polarization of two-mode states of the radiation field. More precisely, we investigate distance-based measures of quantum degrees of polarization, which means that degree of polarization is defined as the shortest distance between the considered state and the set $\mathcal{U}$ of unpolarized states $\hat{\sigma}$ given by Eq. (2). We analyze the Hilbert-Schmidt and Bures distances, which are defined as follows [12]:

$$\mathbb{P}_{\text{HS}}(\hat{\rho}) := \min_{\hat{\sigma}\in\mathcal{U}} \text{Tr}\left[(\hat{\rho}_b - \hat{\sigma})^2\right]; \tag{7}$$

$$\mathbb{P}_{\text{B}}(\hat{\rho}) := 1 - \max_{\hat{\sigma}\in\mathcal{U}} \sqrt{F(\hat{\rho}_b, \hat{\sigma})}, \tag{8}$$

where above we used the third requirement given by Eq. (6). By F we denote the fidelity of two quantum states [22], [23]

$$F(\hat{\rho}, \hat{\zeta}) := \left[\text{Tr}\left(\sqrt{\sqrt{\hat{\rho}}\,\hat{\zeta}\,\sqrt{\hat{\rho}}}\right)\right]^2.$$

In this chapter we use the Hilbert-Schmidt and Bures measures for evaluating the quantum degree of polarization in order to make a comparison to the concurrence of the two-mode state of the quantum radiation field.

3. Three-Photon Bell-Diagonal Mixed States

In this section we investigate a particular family of two-mode mixed states of the quantum radiation field, namely the case of three-photon states; therefore the spanned subspace is of dimension 4. The standard two-mode Fock basis is

$\{|3,0\rangle, |0,3\rangle, |2,1\rangle, |1,2\rangle\}$ from which one can construct the associated Bell basis as follows

$$|\Psi^{\pm}\rangle = \frac{1}{\sqrt{2}}(|3,0\rangle \pm |0,3\rangle);$$
$$|\Phi^{\pm}\rangle = \frac{1}{\sqrt{2}}(|2,1\rangle \pm |1,2\rangle).$$

The states $|\Psi^{+}\rangle = \frac{1}{\sqrt{2}}(|3,0\rangle + |0,3\rangle)$ are the N00N states [24], [25]. On the other hand, the other two states $|\Phi^{\pm}\rangle = \frac{1}{\sqrt{2}}(|2,1\rangle \pm |1,2\rangle)$ were investigated recently in Ref. [26].

We focus our analysis on a relevant class of mixed states, namely the Bell-diagonal states, whose density operator has the following expression

$$\hat{\rho}_{BD} = \alpha|\Phi^{+}\rangle\langle\Phi^{+}| + \beta|\Phi^{-}\rangle\langle\Phi^{-}| + \gamma|\Psi^{+}\rangle\langle\Psi^{+}| + \delta|\Psi^{-}\rangle\langle\Psi^{-}|. \quad (9)$$

The parameters $\alpha, \beta, \gamma, \delta$ satisfy the normalization condition $\alpha+\beta+\gamma+\delta = 1$.

As a measure of entanglement we use the concurrence [27], [15], which in the case of a Bell-diagonal state reads [28]

$$C(\hat{\rho}_{BD}) = \max\{0, \lambda_1 - \lambda_2 - \lambda_3 - \lambda_4\}, \quad (10)$$

where $\{\lambda_1, \lambda_2, \lambda_3, \lambda_4\}$ are the eigenvalues α, β, γ, δ in decreasing order, i.e. $\lambda_1 \geq \lambda_2 \geq \lambda_3 \geq \lambda_4$.

By employing the Eqs. (7) and (8) we obtain in the case of a Bell-diagonal mixed states the following expressions for the quantum degrees of polarization:

a) Hilbert-Schmidt degree of polarization

$$\mathbb{P}_{\mathrm{HS}}(\hat{\rho}_{BD}) = \alpha^2 + \beta^2 + \gamma^2 + \delta^2 - \frac{1}{4}. \quad (11)$$

b) Bures degree of polarization

$$\mathbb{P}_{\mathrm{B}}(\hat{\rho}_{BD}) = 1 - \frac{1}{2}\left(\sqrt{\alpha} + \sqrt{\beta} + \sqrt{\gamma} + \sqrt{\delta}\right). \quad (12)$$

4. Applications

4.1. The Werner State

The Werner state is defined as follows [29]:

$$\hat{\rho}_W = a\,|\Psi^{-}\rangle\langle\Psi^{-}| + (1-a)\frac{1}{4}I, \quad (13)$$

a being a parameter that satisfies $a \in [0, 1]$. The Werner state is a Bell-diagonal state (9) characterized by

$$\begin{aligned} \alpha = \beta = \gamma &= \frac{1-a}{4}; \\ \delta &= \frac{3\,a+1}{4}. \end{aligned} \tag{14}$$

The concurrence of the Werner state is [30]:

$$C(\hat{\rho}_W) = \begin{cases} 0, & \text{if} \quad a \in \left[0, \frac{1}{3}\right] \\ \frac{1}{2}\,(3\,a-1), & \text{if} \quad a \in \left(\frac{1}{3}, 1\right]. \end{cases} \tag{15}$$

The expression of the quantum degrees of polarization is obtained with the help of Eqs. (11) and (12):

$$\begin{aligned} \mathbb{P}_{\rm HS}(\hat{\rho}_W) &= \frac{3}{4}\,a^2; \\ \mathbb{P}_{\rm B}(\hat{\rho}_W) &= 1 - \frac{1}{4}\,\left(\sqrt{1+3\,a} + 3\,\sqrt{1-a}\right). \end{aligned}$$

In Figs. 1 and 2 we plot the concurrence versus the two quantum degrees of polarization: Hilbert-Schmidt and Bures measure, respectively.

There are some certain states for which the concurrence is equal to the quantum degree of polarization. These states are characterized by the parameter $\tilde{a}$ given in Table 1.

Table 1. The particular Werner states with the nice property that the concurrence is equal to the quantum degree of polarization

Concurrence = Degree of polarization	$\tilde{a}$
$C(\hat{\rho}_W) = \mathbb{P}_{\rm HS}(\hat{\rho}_W)$	$\tilde{a} = \frac{6-2\sqrt{3}}{6} \simeq 0.423$
$C(\hat{\rho}_W) = \mathbb{P}_{\rm B}(\hat{\rho}_W)$	$\tilde{a} = 0.35957$

From Figs. 1 and 2 one can notice that for $a < \tilde{a}$, the quantum degree of polarization is greater than the entanglement expressed using the concurrence. For $a > \tilde{a}$, the situation is the opposite: $\mathbb{P}(\hat{\rho}_W) < C(\hat{\rho}_W)$. This means that the concurrence and the quantum degree of polarization are incomparable measures of a two-mode state, i.e. there exist states $\hat{\rho}_{1,W}$ and $\hat{\rho}_{2,W}$ such that $C(\hat{\rho}_{1,W}) < \mathbb{P}(\hat{\rho}_{1,W})$, while $C(\hat{\rho}_{2,W}) > \mathbb{P}(\hat{\rho}_{2,W})$.

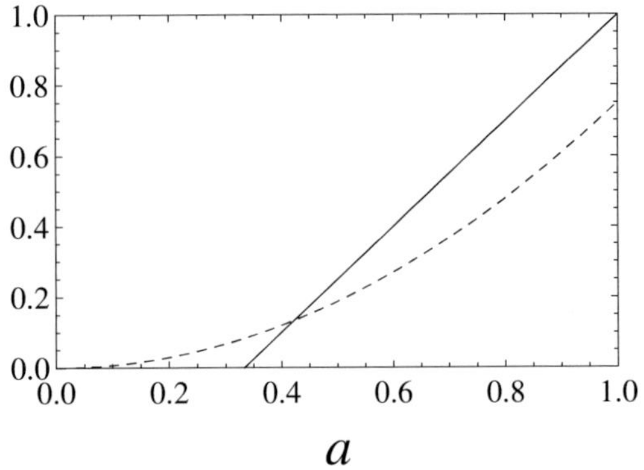

Figure 1. The concurrence (solid) versus the Hilbert-Schmidt degree of polarization (dashed) for the Werner state. The intersection point characterizes a state that verifies the condition $C(\hat{\rho}_W) = \mathbb{P}_{\rm HS}(\hat{\rho}_W)$, this state being defined by $\tilde{a} = \frac{6-2\sqrt{3}}{6}$.

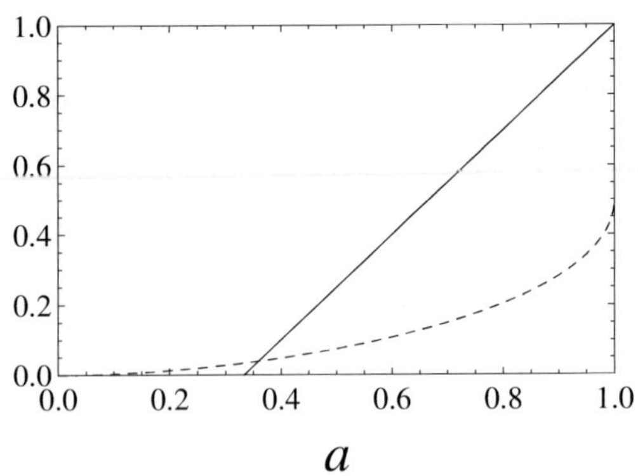

Figure 2. The concurrence (solid) versus the Bures degree of polarization (dashed) for the Werner state. The intersection point defines a state with the condition $C(\hat{\rho}_W) = \mathbb{P}_{\rm B}(\hat{\rho}_W)$, whose parameter is $\tilde{a} = 0.35957$.

4.2. Bell-Diagonal States Depending on a Parameter x

Consider the Bell-diagonal states described by the coefficients α, β, γ, δ that depend on a parameter x as follows:

$$\alpha(x) = \frac{1}{4}(1 - x + x^2 - x^3);$$

$$\begin{aligned}\beta(x) &= \frac{1}{4}(1-x-x^2+x^3);\\ \gamma(x) &= \frac{1}{4}(1+x-x^2-x^3);\\ \delta(x) &= \frac{1}{4}(1+x+x^2+x^3),\end{aligned} \tag{16}$$

with $x \in [-1, 1]$. We denote the density operator of this state by $\hat{\rho}_{BD}(x)$.

The concurrence is obtained by using Eq. (10):

$$\begin{aligned} C(\hat{\rho}_{BD}(x)) &= \begin{cases} \max\{0, \alpha-\beta-\gamma-\delta\}, & \text{if} \quad x \in [-1, 0) \\ \max\{0, \delta-\alpha-\beta-\gamma\}, & \text{if} \quad x \in [0, 1]. \end{cases} \\ &= \begin{cases} \frac{1}{2}(-1-x+x^2-x^3), & \text{if} \quad x \in [-1, -0.544) \\ 0, & \text{if} \quad x \in [-0.544, 0.544] \\ \frac{1}{2}(-1+x+x^2+x^3), & \text{if} \quad x \in (0.544, 1] \end{cases} \end{aligned}$$

Figs. 3 and 4 show the behavior of the concurrence versus the quantum degree of polarization. As in the previous example, we use two distance-type measures for the quantum degree of polarization: Hilbert-Schmidt and Bures.

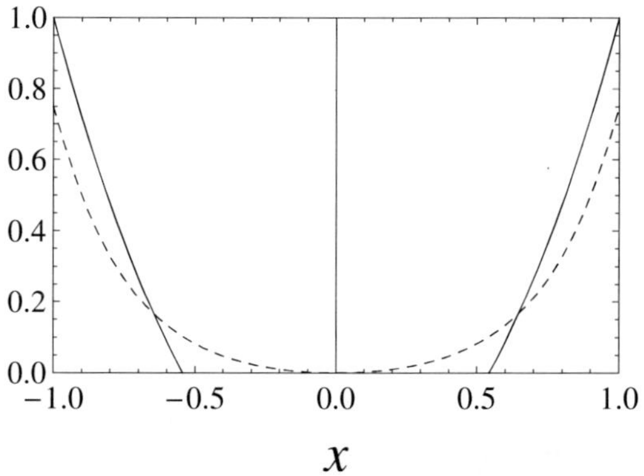

Figure 3. The concurrence (solid) versus the Hilbert-Schmidt degree of polarization (dashed) for the state defined by Eq. (16). The two states characterized by $\tilde{x}_- = -0.645$ and $\tilde{x}_+ = 0.645$ correspond to identical concurrence and Hilbert-Schmidt degree of polarization.

The special states for which the concurrence is equal to the quantum degree of polarization are characterized by the parameters $\tilde{x}_+$ and $\tilde{x}_-$. We show in Table 2 the two values of $\tilde{x}$ for each of the two quantum degrees of polarization.

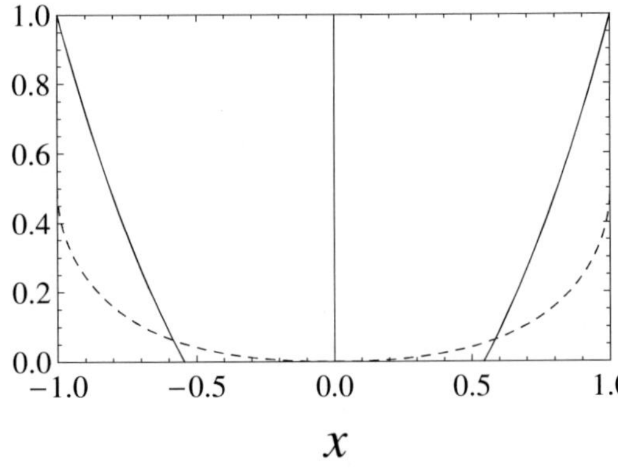

Figure 4. The concurrence (solid) versus the Bures degree of polarization (dashed) for the state defined by Eq. (16). The equality $C(\hat{\rho}_{BD}(x)) = \mathbb{P}_{\mathrm{B}}(\hat{\rho}_{BD}(x))$ gives two solutions $\tilde{x}_- = -0.58437$ and $\tilde{x}_+ = 0.58437$.

Table 2. The particular Bell-diagonal states $\hat{\rho}_{BD}(x)$ of Eq. (16) with the nice property that the concurrence is equal to the quantum degree of polarization

Concurrence = Degree of polarization	$\tilde{x}$
$C(\hat{\rho}_{BD}(x)) = \mathbb{P}_{\mathrm{HS}}(\hat{\rho}_{BD}(x))$	$\tilde{x}_\pm = \pm 0.645$
$C(\hat{\rho}_{BD}(x)) = \mathbb{P}_{\mathrm{B}}(\hat{\rho}_{BD}(x))$	$\tilde{x}_\pm = \pm 0.58437$

Figs. 3 and 4 show an interesting comparison of the entanglement versus the quantum degree of polarization, namely:

$$\mathrm{C}(\hat{\rho}) \ \text{ versus } \ \mathbb{P}(\hat{\rho}) : \begin{cases} C(\hat{\rho}_{BD}(x)) \geq \mathbb{P}(\hat{\rho}_{BD}(x)), & \text{if} \quad x \in [-1, \tilde{x}_-] \\ C(\hat{\rho}_{BD}(x)) < \mathbb{P}(\hat{\rho}_{BD}(x)), & \text{if} \quad x \in (\tilde{x}_-, \tilde{x}_+) \\ C(\hat{\rho}_{BD}(x)) \geq \mathbb{P}(\hat{\rho}_{BD}(x)), & \text{if} \quad x \in [\tilde{x}_+, 1]. \end{cases} \tag{17}$$

In conclusion, there is no dominance relation for the concurrence and the quantum degree of polarization for all the domain of x, i.e. for all the states, which means that the two measures are incomparable.

5. Conclusion

In this chapter we have made a comparison between two quantities that characterize the two-mode states of the radiation field: entanglement and quantum degree of polarization. Our study was focus on three-photon Bell-diagonal mixed states for which we have computed the concurrence (C) and the quantum degree of polarization ($\mathbb{P}$) by using two distance-based measures: Hilbert-Schmidt and Bures. Two examples were considered - the Werner state and Bell-diagonal states depending on a parameter.

The comparison between the concurrence and each of the two quantum degrees of polarization proved to be very useful. The main result of this chapter is that the concurrence and the quantum degree of polarization are incomparable. In other words, there exist states $\hat{\rho}_1$ and $\hat{\rho}_2$ such that $C(\hat{\rho}_1) < \mathbb{P}(\hat{\rho}_1)$, while $C(\hat{\rho}_2) > \mathbb{P}(\hat{\rho}_2)$.

References

[1] Kwiat P. G., Mattle K., Weinfurter H., Zeilinger A., Sergienko A. V., and Shih Y., 1995. New High-Intensity Source of Polarization-Entangled Photon Pairs, *Phys. Rev. Lett.*, 75, 4337.

[2] Bennett C.H., Bessette F., Brassard G., Salvail L., and Smolin J., 1992. Experimental quantum cryptography, *J. Cryptology*, 5, 3.

[3] Muller A., Hertzog T., Huttner B., Tittel W., Zbinden H., and Gisin N., 1997, "Plug and play" systems for quantum cryptography, *Appl. Phys. Lett.*, 70, 793.

[4] Bennett C. H., Brassard G., Crépeau C., Jozsa R., Peres A., and Wootters W. K., 1993. Teleporting an unknown quantum state via dual classical and Einstein-Podolsky-Rosen channels, *Phys. Rev. Lett.*, 70, 1895.

[5] Bouwmeester D., Pan J. W., Mattle M., Eible M., Weinfurther H., and Zeilinger A., 1997. Experimental quantum teleportation, *Nature*, 390, 575.

[6] Bennett C. H. and Wiesner S. J., 1992. Communication via one- and two-particle operators on Einstein-Podolsky-Rosen states, *Phys. Rev. Lett.*, 69, 2881.

[7] Bose S., Vedral V., and Knight P. L., 1998. Multiparticle generalization of entanglement swapping, *Phys. Rev. A*, 57, 822.

[8] Pan J. W., Simon C., Brukner C., and Zeilinger A., 2001. Entanglement purification for quantum communication, *Nature*, 410, 1067.

[9] Joo J, Knight P. L., O'Brien J. L., and Rudolph T. 2007. One-way quantum computation with four-dimensional photonic qudits, *Phys. Rev. A*, 76, 052326.

[10] Mandel L. and Wolf E. 1995. *Optical Coherence and Quantum Optics*, Cambridge University Press, Cambridge, UK, See pp. 342-355.

[11] Simon R., 1990. Nondepolarizing systems and degree of polarization, *Opt. Commun.*, 77, 349.

[12] Klimov A. B., Sánchez-Soto L. L., Yustas E. C., Söderholm J., and Björk G., 2005. Distance-based degrees of polarization for a quantum field, *Phys. Rev. A*, 72, 033813.

[13] Sánchez-Soto L. L., Söderholm J., Yustas E. C., Klimov A. B., and Björk G. 2006, Degrees of polarization for a quantum field, *J. Phys.: Conf. Ser.*, 36, 177.

[14] Björk G, Söderholm J., Sanchez-Soto L. L., Klimov A. B., Ghiu I., Marian P., and Marian T. A., 2010. Quantum degrees of polarization, *Opt. Commun.*, 283, 4440.

[15] Horodecki R., Horodecki P., Horodecki M., and Horodecki K., 2009. Quantum entanglement, *Rev. Mod. Phys.*, 81, 865.

[16] Nogueira K., Silva J. B. R., Goncalves J. R., and Vasconcelos H. M., 2013. Polarization and entanglement of photon-added coherent states, *Phys. Rev. A*, 87, 043821.

[17] Prakash H. and Chandra N., 1971. Density operator of unpolarized radiation, *Phys. Rev. A*, 4, 796.

[18] Prakash H. and Chandra N., 1974. Density operator of unpolarized radiation, *Phys. Rev. A*, 9, 1021.

[19] Agarwal G. S., 1971. On the state of unpolarized radiation, *Lett. Nuovo Cimento*, 1, 53.

[20] Lehner J., Leonhard U., and Paul H., 1996. Unpolarized light: Classical and quantum states, *Phys. Rev. A*, 53, 2727.

[21] Söderholm J., Björk G., and Trifonov A., 2001. Unpolarized light in quantum optics, *Optics and Spectroscopy*, 91, 532.

[22] Jozsa R., 1994. Fidelity for mixed quantum states, *J. Mod. Opt.*, 41, 2315.

[23] Nielsen M. A. and Chuang I. L., 2000, *Quantum Computation and Quantum Information*, Cambridge University Press, Cambridge, UK.

[24] Dowling J. P., 2008. Quantum optical metrology the lowdown on high-N00N states, *Cont. Phys.*, 49, 125.

[25] Shabbir S., Swillo M., and Björk G., 2013. Synthesis of arbitrary, two-mode, high visibility N-photon interference patterns, *Phys. Rev. A*, 87, 053821.

[26] Kowalewska-Kudlaszyk A. and Leonski W., 2014. Nonlinear coupler operating on Werner-like states - entanglement creation, its enhancement and preservation *J. Opt. Soc. Am. B*, 31, 1290.

[27] Wootters W. K., 1998. Entanglement of Formation of an Arbitrary State of Two Qubits, *Phys. Rev. Lett.*, 80, 2245.

[28] Bromley T. R., Cianciaruso M., Lo Franco R., and Adesso G., 2014. Unifying approach to the quantification of bipartite correlations by Bures distance, *J. Phys. A: Math. Theor.*, 47, 405302.

[29] Werner R. F., 1989. Quantum states with Einstein-Podolsky-Rosen correlations admitting a hidden-variable model, *Phys. Rev. A*, 40, 4277.

[30] Ali M., Rau M. R. P., and Alber G., 2010. Quantum discord for two-qubit X states, *Phys. Rev. A*, 81, 042105.

In: New Developments in Quantum Optics ... ISBN: 978-1-63483-394-3
Editor: Nicole Stewart, pp. 113-132

BIBLIOGRAPHY

Advances in slow and fast light, V: 22-24 January 2012, San Francisco, California, United States: Advances in slow and fast light V: 22-24 January 2012, San Francisco, California, United States / Selim M. Shahriar, Frank A. Narducci, editors; sponsored and published by SPIE. Published/Created: Bellingham, Wash.: SPIE, c2012.Description: 1 v. (various pagings): ill.; 28 cm. ISBN: 9780819489166 (pbk.) 0819489166 (pbk.) LC classification: QC446.15 .A3835 2012Related names: Shahriar, Selim M., 1964-Narducci, Frank A. SPIE (Society) Subjects: Quantum optics-Congresses. Nonlinear optics-Congresses. Light-Speed-Congresses. Notes: Includes bibliographical references and index. Series: Proceedings of SPIE, 0277-786X; v. 8273.

From atomic to mesoscale: the role of quantum coherence in systems of various complexities: From atomic to mesoscale: the role of quantum coherence in systems of various complexities / edited by Svetlana A. Malinovskaya (Stevens Institute of Technology, USA), Irina Novikova (College of William and Mary, USA). Published/Produced: Singapore; Hackensack, NJ: World Scientific, [2015] ©2015 Projected pub date: 1506. Description: pages cm, ISBN: 9789814678698 LC classification: QC794.6.C58 F76 2015Related names: Malinovskaya, Svetlana A., editor. Novikova, Irina, 1975- editor. Contents: Quantum magnetism with ultracold molecules / M.L. Wall, K.R.A. Hazzard, and A.M. Rey – Optical manipulation of light scattering in cold atomic Rubidium / R.G. Olave, A.L. Win, K. Kemp, S.J. Roof, S. Balik, M.D. Havey, I.M. Sokolov, and D.V. Kupriyanov -- Seeing spin dynamics in atomic gases / D.M. Stamper-Kurn -- Precision magnetic sensing and imaging using NV-diamond / R.L. Walsworth -- Entanglement and quantum optics with quantum dots / A.P. Burgers, J.R. Schaibley, and D.G. Steel – Enhancement of single-photon sources with metamaterials / M.Y. Shalaginov, S. Bogdanov, V.V. Vorobyov, A. Lagutchev, A.V. Kildishev, A.V. Akimov, A. Boltasseva, and V.M. Shalaev -- Linear optical properties of periodic hybrid materials at oblique incidence: a numerical approach / A. Blake and M. Sukharev – An introduction to Boson-sampling / B.T. Gard, K.R. Motes, J.P. Olson, P.P. Rohde, and J.P. Dowling - New approach to quantum amplification by superradiant emission of radiation / G. Shchedrin, Y. Rostovtsev, X. Zhang, and M.O. Scully - Circularly polarized attosecond pulses and molecular atto-magnetism / A.D. Bandrauk and K.J. Yuan -- Many-electron response of gas-phase fullerene materials to UV and soft X-ray photons / H.S. Chakraborty and M. Magrakvelidze - Collisions and reactions in ultracold gases / N. Balakrishnan and J. Hazra. Subjects: Coherence (Nuclear physics) Quantum systems. Notes: Includes bibliographical references and index.

Laser-based measurements for time and frequency domain applications: a handbook, Maddaloni, Pasquale. Laser-based measurements for time and frequency domain applications: a handbook / Pasquale Maddaloni, Marco Bellini, Paolo De Natale. Published/Created: Boca Raton, FL: Taylor & Francis, c2013.Description: xx, 736 p.: ill.; 27 cm. ISBN: 9781439841518 (hardback) LC classification: QB214 .M34 2012Related names: Bellini, Marco, 1967-De Natale, Paolo. Summary: "Providing a self-contained introductory review of modern laser-based time and frequency measurement techniques, this text represents an interdisciplinary look at the recent developments and future directions of optical frequency meteorology, as well as a range of meteorological disciplines. Suitable for graduate students and practicing physicists and engineers, it discusses the most advanced laser-based spectroscopic measurement techniques, including UV, j1W, visible frequency, and IR laser. The authors, leading optical meteorologists, also cover advanced spectroscopic techniques, experimental quantum optics, and quantum information" - Provided by publisher. Subjects: Time measurements. Frequencies of oscillating systems – Measurement. Lasers--Scientific applications. Optical measurements. Spectrum analysis. Atmosphere--Laser observations. TECHNOLOGY & ENGINEERING / Lasers & Photonics TECHNOLOGY & ENGINEERING / Sensors Notes: Includes bibliographical references (p. 637-700) and index. Series: Series in optics and optoelectronics.

Laser beam propagation: generation and propagation of customized light: Laser beam propagation: generation and propagation of customized light / edited by

Andrew Forbes, CSIR National Laser Centre, Pretoria, South Africa. Published/Produced: Boca Raton: CRC Press, Taylor & Francis Group, [2014] ©2014. Description: xiv, 346 pages: illustrations (some color); 24 cm, ISBN: 9781466554399 (hardcover: acid-free paper) 1466554398 (hardback: acid-free paper) LC classification: QC689.5.L37 L382 2014Related names: Forbes, Andrew, editor of compilation. Contents: Optical waves 3 / Filippus S. Roux - From classical to quantum optics / Thomas Konrad and Sandeep Goyal - Time domain laser beam propagation / Thomas Feurer - Spatial laser beam characterization/Christian Schulze and Daniel Flamm - Characterization of time domain pulses / Thomas Feurer - Generation of laser beams by digital holograms / Jeffrey Davis and Ignacio Morenov - Flat-top beams / Andrew Forbes - Helmholtz-Gauss beams / Julio Gutierrez-Vega - Vector beams / Qiwen Zhan - Low-coherence laser beams / Jari Turunen - The orbital angular momentum of light / Miles Padgett and Reuben Aspden. Subjects: Laser beams. Beam optics. Notes: Includes bibliographical references (pages 323-338) and index.

Laser-induced damage in optical materials: Laser-induced damage in optical materials / edited by Detlev Ristau, Leibniz University of Hannover, Institute for Quantum Optics, and Laser Zentrum Hannover, Department of Laser Components, Hannover, Germany. Published/Produced: Boca Raton: CRC Press, Taylor & Francis Group, [2015] ©2015. Description: xiv, 537 pages: illustrations; 26 cm, ISBN: 9781439872161 (hardcover: alk. paper) 1439872163 (hardcover: alk. paper) LC classification: QC374 .L37 2015. Related names: Ristau, Detlev, editor of compilation. Contents: Laser-induced damage phenomena in optics: an overview / M.J. Soileau – Laser-induced damage (LID) by thermal effects / Roger M. Wood – Defect-induced damage / Semyon Papernov – Self-focusing and nonlinear effects / Vitaly Gruzdev – Femtosecond laser-induced damage in dielectric materials / Luke A. Emmert and Wolfgang Rudolph – Measurement and detection of laser-induced damage/ Jianda Shao – Statistics of laser damage threshold measurements / Jonathan Arenberg – Measurement of light scattering, transmittance, and reflectance / Sven Schröder and Angela Duparré – Quartz and glasses / Laurent Lamaignere – Crystalline materials for UV-applications / Wolfgang Triebel and Christian Mühlig – Materials for lasers: frequency conversion, Q-switching, and active materials / Anne Hildenbrand and Frank R. Wagner – Surface manufacturing and treatment / Jérôme Néauport and Philippe Cormont – Introduction to optical coatings and thin film production / Angus H. Macleod – High-power coatings for NIR lasers / Christopher J. Stolz – Coatings for FS-lasers / Marco Jupé and Detlev Ristau – Spaceborn applications / Wolfgang Riede and Denny

Wernham – Lithography in the deep ultraviolet and extreme ultraviolet / Klaus R. Mann – Free electron lasers / Michelle Shinn - The high-energy petawatt laser phelix and its high-power optics / Stefan Borneis. Subjects: Optical materials--Effect of radiation on. Laser materials--Effect of radiation on. Thin films--Effect of radiation on. Surfaces (Physics)--Effect of radiation on. Notes: Includes bibliographical references and index.

Mathematical optics: classical, quantum, and computational methods: Mathematical optics: classical, quantum, and computational methods / edited by Vasudevan Lakshminarayanan, Maria L. Calvo, Tatiana Alieva. Published/ Produced: Boca Raton: CRC Press, Taylor & Francis Group, [2013] Description: xix, 609 pages: illustrations; 26 cm. ISBN: 9781439869604 (Hardback) LC classification: QC355.3 .M38 2013Related names: Lakshminarayanan, Vasudevan. Calvo, María L. Alieva, Tatiana. Subjects: Optics--Mathematics. Quantum optics. Notes: Includes bibliographical references and index.

Multi-photon quantum information science and technology in integrated optics, Jonathan C.F. Matthews. Matthews, Jonathan C.F. Multi-photon quantum information science and technology in integrated optics / Jonathan C.F. Matthews. Published/Created: Berlin; New York: Springer, c2013. Description: xvii, 117 p.: ill. (some col.); 24 cm. ISBN: 9783642328695 (alk. paper) 3642328695 (alk. paper) LC classification: QC446.2. M388 2013Summary: Photons are an attractive option for testing fundamental quantum physics and developing new quantum-enhanced technology, including highly advanced computers and simulators, as well as precision sensing beyond shot-noise. Traditionally, bulk optical components have been bolted onto optical benches to realize metre-scale quantum circuits. However this approach is ultimately proving unwieldy for increasing the complexity and for scaling up to practical quantum technologies based on photons. The work presented here demonstrates a series of quantum photonic devices based on waveguide circuits embedded in miniature monolithic chips. This represents a paradigm shift in the underlying architecture of quantum optics and provides key building blocks for all-optical and hybrid quantum technologies. - Source other than Library of Congress. Contents: Background and methods - The Hong-Ou-Mandel effect in a waveguide directional coupler - Multi directional-coupler circuit for quantum logic - Quantum interference in a waveguide interferometer - Heralded NOON state generation in waveguide - Two photon quantum walks - Simulating arbitrary quantum statistics with entangled photons - Conclusions and outlook. Subjects: Quantum optics. Multiphoton processes. Multiphoton processes. Quantum optics. Notes: Includes bibliographical references. Series: Springer theses Springer theses.

Optical antennas: Optical antennas / [edited by] Mario Agio, European Laboratory for Nonlinear Spectroscopy (LENS), Florence, Italy; Andrea Alù, University of Texas, Austin, USA. Published/Produced: Cambridge: Cambridge University Press, 2013. Description: xxiii, 455 pages: illustrations; 25 cm. Links: Contributor biographical information http://www.loc.gov/catdir/enhancements/fy1502/2012032866-b.html. Publisher description http://www.loc.gov/catdir/enhancements/fy1502/2012032866-d.html. Table of contents only http://www.loc.gov/catdir/enhancements/fy1502/2012032866-t.html, ISBN: 9781107014145LC classification: TK8360.O65 O68 2013. Related names: Agio, Mario, editor of compilation. Alù, Andrea, editor of compilation. Summary: "This consistent and systematic review of recent advances in optical antenna theory and practice brings together leading experts in the fields of electrical engineering, nano-optics and nano-photonics, physical chemistry and nanofabrication. Fundamental concepts and functionalities relevant to optical antennas are explained, together with key principles for optical antenna modelling, design and characterisation. Recognizing the tremendous potential of this technology, practical applications are also outlined. Presenting a clear translation of the concepts of radio antenna design, near-field optics and field-enhanced spectroscopy into optical antennas, this interdisciplinary book is an indispensable resource for researchers and graduate students in engineering, optics and photonics, physics and chemistry"-- Provided by publisher. Contents: Machine generated contents note: Part I. Fundamentals: 1. From near-field optics to optical antennas D. Pohl; 2. Optical antenna theory, design and applications A. Alù and N. Engheta; 3. Impedance of a nanoantenna F. Marquier and J-J. Greffet; 4. Where high-frequency engineering advances optics: active nanoparticles as nanoantennas R. W. Ziolkowski, S. Arslanagić and J. Geng; 5. Optical antennas for field-enhanced spectroscopy J. Aizpurua and R. Esteban; 6. Directionality, polarization and enhancement by optical antennas T. H. Taminiau, A. Curto and N. F. van Hulst; 7. Antennas, quantum optics and near-field microscopy V. Sandoghdar, M. Agio, X-W. Chen, S. Götzinger and K-G. Lee; 8. Nonlinear optical antennas H. Harutyunyan, G. Volpe and L. Novonty; 9. Coherent control of nano-optical excitations W. Pfeiffer, M. Aeschlimann and T. Brixner; Part II. Modeling, Design and Characterization: 10. Computational electrodynamics for optical antennas O. J. F. Martin; 11. First-principles simulations of near-field effects J. L. Payton, S. M. Morton and L. Jensen; 12. Field distribution near optical antennas at the subnanometer scale C. Pecharromán; 13. Fabrication and optical characterization of nanoantennas J. Prangsma, P. Biagioni and B. Hecht; 14. Probing and imaging of optical nanoantennas with PEEM P. Melchior, D. Bayer and M. Aeschlimann; 15. Fabrication, characterization and applications of optical antenna arrays D. Dregely, J. Dorfmüller, M. Hentschel and H. Giessen; 16. Novel fabrication methods for optical antennas W. Zhou, J. Y. Suh and T. W. Odom; 17. Plasmodic properties of colloidal clusters: towards new nanomaterials and optical circuits J. A. Fan and F. Capasso; Part III. Applications: 18. Optical antennas for information technology and energy harvesting M. L. Brongersma; 19. Nanoantennas for refractive-index sensing T. Shegai, M. Svedendahl, S. Chen, A. Dahlin and M. Ka;ll; 20. Nanoimaging with optical antennas P. Verma and Y. Saito; 21. Aperture optical antennas J. Wenger. Subjects: Optical antennas. Nanophotonics. Technology & Engineering / Electronics / Optoelectronics. Notes: Includes bibliographical references (pages 387-445) and index.

Optical coatings and thermal noise in precision measurement: Optical coatings and thermal noise in precision measurement / edited by Gregory Harry, Timothy P. Bodiya and Riccardo DeSalvo. Published/Created: Cambridge; New York: Cambridge University Press, 2012.Description: xiv, 328 p.: ill.; 26 cm. Links: Cover image http://assets.cambridge.org/97811070/03385/cover/9781107003385.jpg, ISBN: 9781107003385 (hardback) LC classification: TS517.2 .O64 2012. Related names: Harry, Gregory M., 1967- Bodiya, Timothy P. DeSalvo, Riccardo. Summary: "Thermal noise from optical coatings is a growing area of concern and overcoming limits to the sensitivity of high precision measurements by thermal noise is one of the greatest challenges faced by experimental physicists. In this timely book, internationally renowned scientists and engineers examine our current theoretical and experimental understanding. Beginning with the theory of thermal noise in mirrors and substrates, subsequent chapters discuss the technology of depositing coatings and state-of-the-art dielectric coating techniques used in precision measurement. Applications and remedies for noise reduction are also covered. Individual chapters are dedicated to specific fields where coating thermal noise is a particular concern, including the areas of quantum optics/optomechanics, gravitational wave detection, precision timing, high-precision laser stabilisation via optical cavities and cavity quantum electrodynamics. While providing full mathematical detail, the text avoids field-specific jargon, making it a valuable resource for readers with varied backgrounds in modern optics"-- Provided by publisher. Contents: Machine generated contents note: 1. Theory of thermal noise in optical mirrors Y. Levin; 2. Coating technology S. Chao; 3. Compendium of thermal noises in optical mirrors V. B. Braginsky, M. L. Gorodetsky and S. P. Vyatchanin; 4. Coating thermal noise I. Martin and S. Reid; 5. Direct measurements of coating thermal noise K. Numata; 6. Methods of improving thermal noise S. Ballmer and K. Somiya; 7. Substrate thermal noise S. Rowan and I. Martin; 8. Cryogenics K. Numata and K. Yamamoto; 9. Thermo-optic noise M. Evans and G. Ogin; 10. Absorption and thermal issues P. Willems, D. Ottaway and P. Beyersdorf; 11. Optical scatter J. R. Smith and M. E. Zucker; 12. Reflectivity and thickness optimisation I. M. Pinto, M. Principe and R. DeSalvo; 13. Beam shaping A. Freise; 14. Gravitational wave detection D. Ottaway and S. D. Penn; 15. High-precision laser stabilisation via optical cavities M. J. Martin and J. Ye; 16. Quantum optomechanics G. D. Cole and M. Aspelmeyer; 17. Cavity quantum electrodynamics T. E. Northup. Subjects: Optical coatings. Quantum optics. Light--Scattering. Electromagnetic waves--Scattering. SCIENCE / Optics. Notes: Includes bibliographical references.

Optical precursors: from classical waves to single photons: Optical precursors: from classical waves to single photons/J.F. Chen... [et al.]. Published/Created: Singapore; New York: Springer, c2013.Description: vi, 79 p.: ill. (some col.); 23 cm. ISBN: 9789814451932 (pbk.: acid-free paper) 9814451932 (pbk.: acid-free paper)LC classification: QC446.2 .O68 2013Related names: Chen, J. F. (Jiefei) Contents: Theory of optical precursors - Searching for precursors: from microwave to primary optical experiments - Observation of optical precursors in cold atoms -- Optical precursor of a single photon - Discussion and outlook. Subjects: Quantum optics. Quantum optics. Notes: Includes bibliographical references. Series: Springer Briefs in physics, 2191-5423, Springer Briefs in physics.

Physics of quantum devices, Gardiner, C. W. (Crispin W.), 1942: Physics of quantum devices / Crispin Gardiner, University of Otago, New Zealand, Peter Zoller, University of Innsbruck, Austria. Published/Produced: London: Imperial College Press; Hackensack, New Jersey: World Scientific Publishing, 2015. Projected pub date: 1503. Description: page cm. ISBN: 9781783266159 (hardcover: alk. paper) 9781783266166 (pbk.: alk. paper) LC classification: QC446.2 .G369 2015. Related names: Zoller, P. (Peter), 1952-Subjects: Quantum optics--Mathematics. Quantum computers. Computer simulation. Quantum systems. Notes: Includes bibliographical references and index. Series: The quantum world of ultra-cold atoms and light book; II Cold atoms; vol. 4.

Power scaling of enhancement cavities for nonlinear optics, Pupeza, Ioachim, 1980-Power scaling of enhancement cavities for nonlinear optics / Ioachim Pupeza. Published/Created: New York, NY: Springer, 2012. Description: x, 96 p.: ill. (some col.); 24 cm. Links: Contributor biographical information http://www.loc.gov/catdir/enhancements/fy1316/2012938352-b.html. Publisher description http://www.loc.gov/catdir/enhancements/fy1316/2012938352-d.html. Table of contents only http: //www.loc.gov/catdir/enhancements/fy1316/2012938352-t.html. ISBN: 9781461440994 (hbk.: alk. paper)1461440998 (hbk.: alk. paper) LC classification: QC446.2. P868 2012. Subjects: Nonlinear optics. Cavity resonators. Notes: "Doctoral Thesis accepted by The University of Munich/Max Planck Institute for Quantum Optics, Germany." Includes bibliographical references. Dissertation note: Thesis (Ph. D.)--University of Munich. Series: Springer theses, 2190-5053.

Principles of nano-optics, Novotny, Lukas.: Principles of nano-optics / Lukas Novotny, University of Rochester, New York, Bert Hecht, Universität Basel, Switzerland. Edition: 2nd ed. Published/Created: Cambridge: Cambridge University Press, 2012. Description: xvii, 564 p.: ill.; 26 cm. Links: Cover image http://assets.cambridge.org/97811070/05464/cover/9781107005464.jpg, ISBN: 9781107005464 (hardback) LC classification: TA418.9.N35 N68 2012Related names: Hecht, Bert, 1968-Summary: "First published in 2006, this book has become the standard reference on nano-optics. Now in its second edition, the text has been thoroughly updated to take into account new developments and research directions. While the overall structure and pedagogical style of the book remain unchanged, all existing chapters have been expanded and a new chapter has been added. Adopting a broad perspective, the authors provide a detailed overview of the theoretical and experimental concepts that are needed to understand and work in nano-optics, across subfields ranging from quantum optics to biophysics. New topics of discussion include: optical antennas; new imaging techniques; Fano interference and strong coupling; reciprocity; metamaterials; and cavity optomechanics. With numerous end-of-chapter problem sets and illustrative material to expand on ideas discussed in the main text, this is an ideal textbook for graduate students entering the field. It is also a valuable reference for researchers and course teachers"-- Provided by publisher."The introduction of new concepts and applications. Applications of near-field optics soon covered a large span ranging from fundamental physics and materials science to biology and medicine. Following a logical development, the strong interest in near-field optics gave birth to the fields of singlemolecule spectroscopy and plasmonics, and inspired new theoretical work associated with the nature of optical near-fields. In parallel, relying on the momentum of the flowering nanosciences, researchers started to tailor nanomaterials with novel optical properties. Photonic crystals, single-photon sources and optical microcavities are products of this effort"--Provided by publisher. Contents: Machine generated contents note: Preface; 1. Introduction; 2. Theoretical foundations; 3. Propagation and focusing of optical fields; 4. Resolution and localization; 5. Nanoscale optical microscopy; 6. Near-field optical probes; 7. Probe-sample distance control; 8. Optical interactions; 9. Quantum emitters; 10. Dipole emission near planar interfaces; 11. Photonic crystals, resonators, and cavity optomechanics; 12. Surface plasmons; 13. Optical antennas; 14. Forces in confined fields; 15. Fluctuation-induced interactions; 16. Theoretical methods in nano-optics; Appendices; Index. Subjects: Nanostructured materials. Near-field microscopy. Quantum optics. Photonics. Nanophotonics. SCIENCE / Optics. Dewey class no.: 535/.15.

Quantum communications and quantum imaging X: Quantum communications and quantum imaging: Quantum communications and quantum imaging X: Quantum communications and quantum imaging / Ronald E. Meyers, Yanhua Shih, Keith S. Deacon, editors; sponsored and published by SPIE. Published/Created: Bellingham: Intl Soc Of Optical Eng, 2012. Description: 1 volume (various pagings): illustrations; 28 cm. ISBN: 0819492353 (pbk.) 9780819492357 (pbk.) Related names: Meyers, Ronald E. Shih, Yanhua. Deacon, Keith S.SPIE (Society) Subjects: Optical communications-Congresses. Quantum optics-Congresses. Quantum electronics-Congresses. Notes: Includes bibliographical references and index. Series: Proceedings of SPIE, 0277-786X; v. 8518.

Quantum communications and quantum imaging XII: 18, 20-21 August 2014, San Diego, California, United States: Quantum communications and quantum imaging XII: 18, 20-21 August 2014, San Diego, California, United States / Ronald E. Meyers, Yanhua Shih, Keith S. Deacon, editors; sponsored ... by SPIE. Published/Produced: Bellingham, Washington: SPIE, [2014] ©2014 Description: 1 volume (various pagings): illustrations; 28 cm. ISBN: 9781628412529Related names: Meyers, Ronald E. Shih, Yanhua.

Deacon, Keith S.SPIE (Society) Subjects: Optical communications-Congresses. Quantum optics-Congresses. Quantum electronics-Congresses. Subject keywords: Quantum communications Quantum imaging, Notes: Includes bibliographical references and index. Series: Proceedings of SPIE, 0277-786X; volume 9225Proceedings of SPIE--the International Society for Optical Engineering; v. 9225. 0277-786X.

Quantum control of molecular processes, Shapiro, Moshe.: Quantum control of molecular processes / Moshe Shapiro, Paul Brumer. Edition: 2nd, rev. and enl. ed. Published/Created: Weinheim: Wiley-VCH, c2012.Description: xvii, 544 p.: ill.; 24 cm. Links: Inhaltsverzeichnis. http://bvbr.bib-bvb.de:8991/F?func=service&doc_library=BVB01&doc_number=024625246&line_number=0001&func_code=DB_RECORDS&service_type=MEDIA, http://deposit.d-nb.de/cgi-bin/dokserv?id=3847397&prov=M&dok_var=1&dok_ext=htm Inhaltstext ISBN: 9783527409044 (print: acid-free paper) 3527409041 (print: acid-free paper) LC classification: QC446.2 .S53 2012. Related names: Brumer, Paul. Shapiro, Moshe. Principles of the quantum control of molecular processes. Contents: Preliminaries of the interaction of light with matter - Weak-field photo dissociation - Weak-field coherent control - Control of intramolecular dynamics - Optimal control theory - Decoherence and its effects on control - Case studies in coherent control - Coherent control of bimolecular processes - The interaction of light with matter: a closer look - Coherent control with quantum light - Coherent control beyond the weak-field regime: bound states and resonances - Photodissociation beyond the weak-field regime - Coherent control beyond the weak-field regime: the continuum - Coherent control of the synthesis and purification of chiral molecules - Strong-field coherent control - Coherent control with few-cycle pulses - Case studies in optimal control - Coherent control in the classical limit. Subjects: Quantum optics. Coherence (Optics) Molecular dynamics. Molekülphysik. Quantenoptik. Notes: First ed. published as: Principles of the quantum control of molecular processes. Hoboken, N.J.: Wiley-Interscience, c2003Includes bibliographical references (p. 513-535) and index. Dewey class no.: 535/.15.

Quantum gases: finite temperatures and non-equilibrium dynamics: Quantum gases: finite temperatures and non-equilibrium dynamics / edited by Nick Proukakis ... [et al.]. Published/Created: London: Imperial College Press; Singapore; Hackensack, NJ: Distributed by World Scientific, c2013. Description: xxiv, 554 p.: ill.; 24 cm. ISBN: 9781848168107 (hbk.)

1848168101 (hbk.) LC classification: QC175.16.P5 Q36 2013. Related names: Proukakis, Nick. Summary: A broad overview of the principal theoretical techniques applied to non-equilibrium and finite temperature quantum gases. Covering Bose-Einstein condensates, degenerate Fermi gases, and the more recently realised exciton-polariton condensates, it fills a gap by linking between different methods with origins in condensed matter physics, quantum field theory, quantum optics, atomic physics and statistical mechanics - Source other than Library of Congress. Subjects: Cold gases. Quantum theory. Notes: Includes bibliographical references (p. 461-537) and indexes. Series: Cold atoms; v. 1Cold atoms; v. 1.

Quantum optics, Agarwal, G. S. (Girish S.), 1946-Quantum optics / Girish S. Agarwal, Oklahoma State University. Published/Produced: Cambridge, UK: Cambridge University Press, 2013.Description: xvi, 491 pages: illustrations; 26 cm. Links: Cover image http://assets.cambridge.org/97811070/06409/cover/9781107006409.jpg, Contributor biographical information http://www.loc.gov/catdir/enhancements/fy1211/2012025069-b.html, Publisher description http://www.loc.gov/catdir/enhancements/fy1211/2012025069-d.html, Table of contents only http://www.loc.gov/catdir/enhancements/fy1211/2012025069-t.html, ISBN: 9781107006409 (hardback) LC classification: QC446.2. A33, 2013 Summary: "In the last decade many important advances have taken place in the field of quantum optics, with numerous potential applications. Ideal for graduate courses on quantum optics, this textbook provides an up-to-date account of the basic principles of the subject. Focusing on applications of quantum optics, the textbook covers recent developments such as engineering of quantum states, quantum optics on a chip, nano-mechanical mirrors, quantum entanglement, quantum metrology, spin squeezing, control of decoherence and many other key topics. Readers are guided through the principles of quantum optics and their uses in a wide variety of areas including quantum information science and quantum mechanics. The textbook features end-of-chapter exercises with solutions available for instructors at www.cambridge.org/9781107006409. It is invaluable to both graduate students and researchers in physics and photonics, quantum information science and quantum communications" - Provided by publisher. Contents: Machine generated contents note: 1. Quantized electromagnetic field and coherent state representations; 2. Nonclassicality of radiation fields; 3. Two mode squeezed states and quantum entanglement; 4. Non-Gaussian nonclassical states; 5. Optical interferometry with single photons and nonclassical light; 6. Polarization and orbital angular momentum of quantum fields; 7. Absorption, emission, and scattering of radiation; 8. Partial coherence in multimode quantum fields; 9. Open quantum systems; 10. Amplification and attenuation of quantum fields; 11. Quantum coherence, interference and squeezing in two-level systems; 12. Cavity quantum electrodynamics; 13. Absorption, emission and scattering from two-level atoms; 14. Quantum interference and entanglement in radiating systems; 15. Near-field radiative effects; 16. Decoherence and disentanglement in two-level systems; 17. Coherent control of the optical properties; 18. Dispersion management and ultraslow light; 19. Single photons and nonclassical light in integrated structures; 20. Quantum optical effects in nano-mechanical systems; Index. Subjects: Quantum optics--Textbooks. SCIENCE / Optics. Notes: Includes bibliographical references and index. Dewey class no.: 535/.15

Quantum optics II: 16-17 April 2012, Brussels, Belgium.: Quantum optics II: 16-17 April 2012, Brussels, Belgium. Published/Created: Bellingham, Washington: SPIE, [2012] Description: 1 volume (various pagings): illustrations; 28 cm. ISBN: 97808194913290819491322Related names: SPIE (Society) Notes: Includes bibliographical references and author index. Series: Proceedings of SPIE, 0277-786X; volume 8440.

Quantum optics for engineers, Duarte, F. J. (Frank J.): Quantum optics for engineers / F.J. Duarte. Published/Produced: Boca Raton, [Florida]: Taylor & Francis, [2014] Description: xxxi, 411 pages: illustrations; 25 cm. ISBN: 9781439888537 (hbk.: alk. paper) LC classification: QC446.2. D83 2014Subjects: Quantum optics. Notes: Includes bibliographical references and index.

Quantum optics with semiconductor nanostructures: Quantum optics with semiconductor nanostructures / edited by Frank Jahnke. Published/ Produced: Oxford: WP, Woodhead Publishing, 2012. Description: xxiv, 577 pages: illustrations; 24 cm. ISBN: 9780857092328 (print) 0857092324 (print) LC classification: QC446.2 .Q38 2012. Related names: Jahnke, Frank, editor. Contents: pt. 1. Single quantum dot systems; Resonance fluorescence emission from single semiconductor quantum dots coupled to high-quality microcavities / S.M. Ulrich, A. Ulhaq and P. Michler, University of Stuttgart, Germany; Quantum optics with single quantum dots in photonic crystal cavities / A. Majumdar, M. Bajcsky, K. Rivoire, S. Buckley, A. Faraon, E.D. Kim, D. Englund, J. Vučkovi,

Stanford University, USA; Modelling single quantum dots in microcavities / C. Gies, M. Florian and F. Jahnke, University of Bremen, Germany and P. Gartner, University of Bremen, Germany and National Institute of Materials Physics, Bucharest-Magurele, Romania - pt. 2. Nanolasers with quantum dot emitters; Highly efficient quantum dot micropillar lasers / S. Reitzenstein, Technical University Berlin, Germany and A. Forchel, University of Wúrzburg, Germany; Photon correlations in semiconductor nanostructures / M. Assmann and M. Bayer, Technische Universität Dortmund, Germany; Emission properties of photonic crystal nanolasers / S. Strauf, Stevens Institute of Technology, USA; Deformed wavelength-scale microdisk lasers with quantum dot emitters / J-B. Shim, A. Eberspächer and J. Wiersig, Universität Magdeburg, Germany, J. Unterhinninghofen, OEC AG, Germany, Q.H. Song, Harbin Institute for Technology, China, L. Ge, Princeton University, USA, H. Cao and A.D. Stone, Yale University, USA - pt. 3. Light-matter interaction in semiconductor nanostructures; Photon statistics and entanglement in phonon-assisted quantum light emission from semiconductor quantum dots / A. Carmele, M-R. Dachner, J. Kabuss, M. Richter, F. Milde and A. Knorr, Technical University Berlin, Germany; Luminescence spectra of quantum dots in microcavities / F.P. Laussy, Walter Schottky Institut, Germany, E. Del Valle, TU München, Germany, A Laucht, Walter Schottky Institut, Germany, A. Gonzalez-Tudela, Universidad Autónoma de Madrid, Spain, M. Kaniber and J.J. Finley, Walter Schottky Institut, Germany and C. Tejedor, Universidad Autónoma de Madrid, Spain; Photoluminescence from a quantum-dot-cavity system / G. Tarel and V. Savona, École Polytechnique Fédérale de Lausanne (EPFL), Switzerland, M. Winger, T. Volz and A. Imamoglu, Eidgenössische Technische Hochschule Zürich (ETHZ), Switzerland; Quantum optics with quantum-dot and quantum-well systems / L. Schneebeli, University of Arizona, USA, M. Kira and S.W. Koch, Philipps-Universität Marburg, Germany -- pt. 4. Semiconductor cavity quantum electrodynamics (QED); All-solid-state quantum optics employing quantum dots in photonic crystals / P. Lodahl, University of Copenhagen, Denmark; One-dimensional photonic crystal nanobeam cavities / J. Hendrickson, Air Force Research Laboratory, USA, A. Homyk and A. Scherer, California Institute of Technology, USA, T. Alasaarela, A. Säynätjoki and S. Honkanen, Aalto University School of Electrical Engineering, Finland, B.C. Richards,

Emcore Photovoltaics, USA, J-Y. Kim and Y-H. Lee, Korea Advanced Institute of Science and Technology, Korea, R. Gibson, M. Gehl, J.D. Olitzky, S. Zandbergen, H.M. Gibbs and G. Khitrova, University of Arizona, USA; Growth of II-VI and III-nitride quantum-dot microcavity systems / C. Kruse, S. Figge and D. Hommel, University of Bremen, Germany -- pt. 5. Ultrafast phenomena; Femtosecond quantum optics with semiconductor nanostructures / R. Bratschitsch, Chemnitz University of Technology, Germany, R. Huber, University of Regensburg, Germany and A. Leitenstorfer, University of Konstanz, Germany; Coherent optoelectronics with quantum dots / S. Michaelis de Vasconcellos, S. Gordon, D. Mantei, Y.A. Leier, M. al-Hmoud, W. Quiring and A. Zrenner, Universität Paderborn, Germany. Subjects: Quantum optics. Nonlinear optics. Nanostructures--Optical properties. Nanostructures--Optical properties. Nonlinear optics. Quantum optics. Notes: Includes bibliographical references and index. Series: Woodhead Publishing series in electronic and optical materials, 2050-1501; number 28Woodhead Publishing series in electronic and optical materials; no. 28.Dewey class no.: 535.15.

Quantum opto-mechanics with micromirrors: combining nano-mechanics with quantum optics: doctoral thesis accepted by University of Vienna, Austria Gröblacher, Simon.: Quantum opto-mechanics with micromirrors: combining nano-mechanics with quantum optics: doctoral thesis accepted by University of Vienna, Austria / Simon Gröblacher. Published/ Produced: Heidelberg; New York: Springer, [2013] ©2012. Description: xii, 142 pages: illustrations (some color); 24 cm. ISBN: 9783642349546 (alk. paper) 3642349544 (alk. paper) LC classification: QC446.2. G76 2013. Subjects: Quantum optics. Quantum optics. Notes: Includes bibliographical references. Series: Springer theses: recognizing outstanding Ph. D. research, 2190-5053.Springer theses.

Quantum paths: festschrift in honor of Berge Englert on his 60th birthday: Quantum paths: festschrift in honor of Berge Englert on his 60th birthday/ [edited by] Hui Khoon Ng (Yale-NUS College, Singapore), Rui Han (Centre for Quantum Technologies, Singapore). Published/Produced: Hackensack, New Jersey: World Scientific, [2015] ©2015. Projected pub date: 1505 Description: pages cm. ISBN: 9789814651837 (hard cover: alk. paper) LC classification: QC173.97 .Q36 2015Related names: Ng, Hui Khoon, editor. Han, Rui, 1985 or 1986- editor. Englert, Berthold-

Georg, 1953- honouree. Contents: Quantum optics and open quantum systems - Wave-particle duality - Mutually unbiased bases and the mean king's problem - Foundations of quantum mechanics - Cold atoms - Information about quantum systems - Personal anecdotes. Subjects: Englert, Berthold-Georg, 1953- honouree. Quantum theory.

Quantum sensing and nanophotonic devices IX: 22-26 January 2012, San Francisco, California, United States: Quantum sensing and nanophotonic devices IX: 22-26 January 2012, San Francisco, California, United States / Manijeh Razeghi, Eric Tournie, Gail J. Brown, editors; sponsored and published by SPIE. Published/Created: Bellingham, Wash.: SPIE, c2012. Description: 1 v. (various pagings): ill.; 28 cm. ISBN: 9780819489110819489115 LC classification: QC688 .Q36 2012QC371. S67 v. 8268 Related names: Razeghi, M. Tournié, Eric, 1962-Brown, Gail J. SPIE (Society) Subjects: Quantum electronics--Congresses. Quantum optics--Congresses. Nanostructured materials--Optical properties--Congresses. Notes: Includes bibliographical references and index. Series: Proceedings of SPIE, 0277-786X; v. 8268Proceedings of SPIE--the International Society for Optical Engineering; v. 8268. 0277-786X.

Semiconductor quantum optics, Kira, Mackillo, 1969-Semiconductor quantum optics / Mackillo Kira and Stephan W. Koch. Published/Created: Cambridge; New York: Cambridge University Press, 2012.Description: xiii, 643 p.: ill.; 26 cm. ISBN: 9780521875097 (hardback) LC classification: QC611.6.Q36 K57 2012. Related names: Koch, S. W. (Stephan W.) Summary: "The emerging field of semiconductor quantum optics combines semiconductor physics and quantum optics, with the aim of developing quantum devices with unprecedented performance. In this book researchers and graduate students alike will reach a new level of understanding to begin conducting state-of-the-art investigations. The book combines theoretical methods from quantum optics and solid-state physics to give a consistent microscopic description of light-matter- and many-body-interaction effects in low-dimensional semiconductor nanostructures. It develops the systematic theory needed to treat semiconductor quantum-optical effects, such as strong light-matter coupling, light-matter entanglement, squeezing, as well as quantum-optical semiconductor spectroscopy. Detailed derivations of key equations help readers learn the techniques and nearly 300 exercises help test their understanding of the materials covered. The book is accompanied by a website hosted by the authors, containing further discussions on topical issues, latest trends and publications on the field. The link can be found at www.cambridge.org/9780521875097 "- Provided by publisher." The emerging field of semiconductor quantum optics combines semiconductor physics and quantum optics, with the aim of developing quantum devices with unprecedented performance. In this book researchers and graduate students alike will reach a new level of understanding to begin conducting state-of-the-art investigations. The book combines theoretical methods from quantum optics and solid-state physics to give a consistent microscopic description of light-matter- and many-body-interaction effects in low dimensional semiconductor nanostructures. It develops the systematic theory needed to treat semiconductor quantum-optical effects, such as strong light-matter coupling, light-matter entanglement, squeezing, as well as quantum-optical semiconductor spectroscopy. Detailed derivations of key equations help readers learn the techniques and nearly 300 exercises help test their understanding of the materials covered"-- Provided by publisher. Contents: Machine generated contents note: 1. Central concepts in classical mechanics; 2. Central concepts of classical electrodynamics; 3. Central concepts in quantum mechanics; 4. Central concepts in stationary quantum theory; 5. Central concepts in measurement theory; 6. Wigner's phase-space representation; 7. Hamiltonian formulation of classical electrodynamics; 8. System Hamiltonian of classical electrodynamics; 9. System Hamiltonian in the generalized Coulomb gauge; 10. Quantization of light and matter; 11. Quasiparticles in semiconductors; 12. Band structure of solids; 13. Interactions in semiconductors; 14. Generic quantum dynamics; 15. Cluster-expansion representation of the quantum dynamics; 16. Simple many-body systems; 17. Hierarchy problem for dipole systems; 18. Two-level approximation for optical transition; 19. Self-consistent extension of the two-level approach; 20. Dissipative extension of the two-level approach; 21. Quantum-optical extension of the two-level approach; 22. Quantum dynamics of two-level system; 23. Spectroscopy and quantum-optical correlations; 24. General aspects of semiconductor optics; 25. Introductory semiconductor optics; 26. Maxwell-semiconductor Bloch equations; 27. Coherent vs. incoherent excitons; 28. Semiconductor luminescence equations; 29. Many-body aspects of the semiconductor luminescence; 30. Advanced semiconductor quantum optics; Appendix;

Index. Subjects: Semiconductors. Quantum optics. Quantum electrodynamics. SCIENCE / Physics Notes: Includes bibliographical reference and index.

Strong light-matter coupling: from atoms to solid-state systems: Strong light-matter coupling: from atoms to solid-state systems / editors; Alexia Auffèves, Institut Néel-CNRS, France, Dario Gerace, Università di Pavia, Italy, Maxime Richard, Institut Néel-CNRS, France, Stefano Portolan, Institut Néel-CNRS, France, Marcelo França Santos, Universidade Federal de Minas Gerais, Brazil, Leong Chuan Kwek, National University of Singapore and Nanyang Technological University, Singapore, Christian Miniatura, INLN-CNRS, University of Nice Sophia, France, CQT, National University of Singapore, Singapore. Published/Produced: New Jersey: World Scientific, [2014] ©2014. Description: ix, 292 pages: illustrations; 26 cm. ISBN: 9789814460347 (hbk.: alk. paper) 9814460346 (hbk.: alk. paper) LC classification: QC446.2 .S773 2014. Related names: Auffèves, Alexia, editor. Gerace, Dario, editor. Richard, Maxime, 1977- editor. Portolan, Stefano, editor. Santos, Marcelo França, editor. Kwek, Leong Chuan, editor. Miniatura, C. (Christian), editor. Contents: Cavity QED in atomic physics / Serge Haroche and Jean-Michel Raimond -- Exciton-polaritons in bulk semiconductors and in confined electron and photon systems / Lucio Claudio Andreani -- Experimental circuit QED / Patrice Bertet -- Quantum open systems / H.J. Carmichael -- Basic concepts in quantum information / Steven M. Girvin -- Cavity polaritons: crossroad between non-linear optics and atomic condensates / Alberto Amo and Jacqueline Bloch -- Quantum plasmonics / Darrick Chang -- Quantum polaritonics / S. Portolan, O. Di Stefano and S. Savasta -- Optical signal processing with enhanced nonlinearity in photonic crystals / A. De Rossi and S. Combrí. Subjects: Quantum optics. Quantum electrodynamics. Notes: Includes bibliographical references.

The angular momentum of light: The angular momentum of light / edited by David L. Andrews, University of East Anglia, Mohamed Babiker, The University of York. Published/Created: Cambridge, UK: Cambridge University Press, 2013. Description: xvi, 425 pages: illustrations; 26 cm. Links: Cover image http://assets.cambridge.org/97811070/06348/cover/9781107006348.jpg. Contributor biographical information http://www.loc.gov/catdir/enhancements/fy1205/2011052383-b.html. Publisher description http://www.loc.gov/catdir/enhancements/fy1205/

2011052383-d.html. Table of contents only http://www.loc.gov/catdir/enhancements/fy1205/2011052383-t.html. ISBN: 9781107006348 (hardback) LC classification: QC446.3.A54 A54 2013Related names: Andrews, David L., 1952-Babiker, Mohamed. Summary: "Recent developments in the angular momentum of light present fresh challenges to long established concepts and pave the way for new and wide-ranging applications. The scope for structured light such as optical vortices, in particular, now extends from microfluidics to quantum information. This is the first comprehensive edited collection dealing with light carrying spin and orbital angular momentum, covering both fundamental and applied aspects. Written by internationally leading specialists, the chapters have been compiled to reflect the latest scientific progress and to address the multitude of theoretical, experimental and technical issues associated with this vibrant and exciting field. The volume is an authoritative reference for academic researchers and graduate students engaged in theoretical or experimental study of optical angular momentum and its applications. It will also benefit professionals in physics, optics and optical engineering, chemistry and biology"-- Provided by publisher. Contents: Machine generated contents note: Preface D. L. Andrews and M. Babiker; 1. Light beams carrying orbital angular momentum J. B. Götte and S. M. Barnett; 2. Vortex transformation and vortex dynamics in optical fields G. Molina-Terriza; 3. Vector beams in free space E. J. Galvez; 4. Optical beams with orbital angular momentum in nonlinear media A. S. Desyatnikov and Y. S. Kivshar; 5. Ray optics, wave optics and quantum mechanics G. Nienhuis; 6. Quantum formulation of angle and orbital angular momentum J. B. Götte and S. M. Barnett; 7. Dynamic rotational frequency shift I. Bialynicki-Birula and Z. Bialynicka-Birula; 8. Spin-orbit interactions of light in isotropic media K. Y. Bliokh, A. Aiello and M. A. Alonso; 9. Quantum electrodynamics, angular momentum and chirality D. L. Andrews and M. Babiker; 10. Trapping of charged particles by Bessel beams I. Bialynicki-Birula, Z. Bialynicka-Birula and N. Drozd; 11. Theory of atoms in twisted light M. Babiker, D. L. Andrews and V. E. Lembessis; 12. An experimentalist's introduction to orbital angular momentum for quantum optics J. Romero, D. Giovannini, S. Franke-Arnold and M. J. Padgett; 13. Measurement of light's orbital angular momentum M. P. J. Lavery, J. Courtial and M. J. Padgett; 14. Efficient generation of optical twisters using helico-conical beams V. R. Daria, D.

Palima and J. Glückstad; 15. Self similar modes of coherent diffusion with orbital angular momentum O. Firstenberg, M. Shuker, R. Pugatch and N. Davidson; 16. Dimcnsionality of azimuthal entanglement M. van Exter, E. Eliel and H. Woerdman; Index. Subjects: Angular momentum. Quantum optics. SCIENCE / Optics. Notes: Includes bibliographical references and index.

The quantum theory of nonlinear optics, Drummond, P. D. (Peter D.), 1950: The quantum theory of nonlinear optics / Peter D. Drummond, Swinburne University of Technology, Melbourne, Mark Hillery, Hunter College, City University of New York. Published/Produced: Cambridge; New York: Cambridge University Press, 2014.Description: viii, 373 pages: illustrations; 26 cm. ISBN: 9781107004214 (hardback) 1107004217 (hardback) LC classification: QC446.2 .D78 2014Related names: Hillery, Mark Stephen. Subjects: Nonlinear optics. Quantum optics. Quantum theory. Notes: Includes bibliographical references and index.

The quantum world of ultra-cold atoms and light, Gardiner, C. W. (Crispin W.), 1942- author: The quantum world of ultra-cold atoms and light / Crispin Gardiner, University of Otago, New Zealand, Peter Zoller, University of Innsbruck, Austria. Published/Produced: London: Imperial College Press, 2014-Description: volumes: illustrations (some color); 24 cm. ISBN: 1783264608978178326460517832646169781783264612 (pbk) LC classification: QC446.2 .G372 2014. Related names: Zoller, P. (Peter), 1952- author. Incomplete contents: Book 1. Foundations of quantum optics. Subjects: Quantum optics. Quantum optics. Notes: Conceived as a trilogy. Includes bibliographical references (pages 293-296) and indexes. Series: Cold atoms, 2045-9734; volume 2, Cold atoms; v. 2.

Ultracold atoms in optical lattices: simulating quantum many-body systems, Lewenstein, Maciej, 1955-:Ultracold atoms in optical lattices: simulating quantum many-body systems / Maciej Lewenstein, Anna Sanpera and Verònica Ahufiniger. Edition: 1st ed. Published/Created: Oxford, U.K.: Oxford University Press, 2012. Description: xiv, 479 p.: ill.; 26 cm. Links: Contributor biographical information http://www.loc.gov/catdir/enhancements/fy1307/2012932694-b.html. Publisher description http://www.loc.gov/catdir/enhancements/fy1307/2012932694-d.html
Table of contents only http://www.loc.gov/catdir/enhancements/fy1307/2012932694-t.html ISBN: 97801995731270199573123LC classification:

QC446.2 .L48 2012 Portion of title: Simulating quantum many-body systems Related names: Sanpera, Anna Ahufinger, Verònica. Contents: 1. Introduction - 2. Statistical physics of condensed matter: basic concepts - 3. Ultracold gases in optical lattices: basic concepts - 4. Quantum simulators of condensed matter - 5. Bose-Hubbard models: methods of treatment - 6. Fermi and Fermi-Bose Hubbard models: methods of treatment - 7. Ultracold spinor atomic gases - 8. Ultracold dipolar gases - 9. Disordered ultracold atomic gases - 10. Frustrated ultracold atom systems - 11. Ultracold atomic gases in 'artificial' gauge fields - 12. Many-body physics from a quantum information perspective - 13. Quantum information with lattice gases - 14. Detection of quantum systems realized with ultracold atoms - 15. Perspectives: beyond standard optical lattices. Subjects: Quantum optics. Optical lattices. Many-body problem. Quantum theory. Notes: Includes bibliographical references (p. [439]-470) and index.

INDEX

A

B

C

D

M

N

O

P

Q

R

S

T

V

W

X